The Beekeepers Annual 2014

THE BEEKEEPERS ANNUAL
IS PUBLISHED BY
NORTHERN BEE BOOKS
MYTHOLMROYD,
WEST YORKSHIRE

PRINTED BY
LIGHTNING SOURCE, UK
ISBN 978-1-908904-47-8

MMXIII

EDITOR, JOHN PHIPPS
NEOCHORI, 24024 AGIOS NIKOLAOS,
MESSINIAS, GREECE
EMAIL manifest@runbox.com

SET IN HELVETICA LT BY D&P Design and Print

The
Beekeepers
Annual 2014

NB

Front Cover

"Praise of the Bees"
Barberini Exultet Scroll, circa. 1087

The legend of the Virgin reproduction of bees, which began with Virgil, was taken by St. Ambrose and was widely distributed by the deacons in their poetic compositions, and was later introduced into the official liturgy of the Church and well fed for centuries the piety of the faithful during Easter prayers. The bee was the most perfect symbol of virginal purity along with images and figures of the Virgin Mary. Indeed in ancient times bees must have been well respected, as their industriousness even makes it into the Bible, or at least the product of it, as *"a land flowing with milk and honey"* (Exodus 3:8). Bees were also included in the Exultet, - a song of prayer sung during the Easter Vigil.
The part of the prayer goes:

"Accept this Easter candle, a flame divided but undimmed, a pillar of fire that glows to the honor of God.

For it is fed by the melting wax, which the mother bee brought forth to make this precious candle."

The Exultet or Easter Proclamation hymn was probably composed between the fifth and seventh century AD. It is used primarily in Western Christianity. The Exultet is the hymn of praise, sung ideally by the deacon, before the paschal candle during the Easter Vigil in the Roman Rite of Mass.

The song of the Exultet was traditionally written in a book, which was read in its length (unlike the ancient way). The Deacon dangling in front of the pulpit already read the text, and illustrations, made "upside down" allowed the faithful to follow along in song. During the dark ages only the wealthy and the clergy could read, so the illuminated pictorials in the Exultet were an essential part of the religious experience. Though it was dark, the church is lit by candle and it was therefore part of a symbolic element.

The Barberini Exultet Scroll pictured above is very interesting because more than just hives and bees are depicted. It has a fairly complete picture of beekeeping, which shows an overview of beekeeping - with bees everywhere foraging flowers. A honeycomb is being cut from the hive, harvested with a sickle knife, aided by a young assistant at his side. On the right, two men are responsible for collecting a swarm. One man cuts the branch on which the swarm is hanging, the other holds in one hand a basic smoker to calm the swarm and place it in a new horizontal hive (which is vertical), held in the other hand.

Over the following centuries, Exultet scrolls were reproduced by hand, so the illuminated illustrations depicting bees would vary with each scroll. During the twelfth century the legend of the virgin reproduction of bees began to fall into disuse, and during the liturgical reform of Pope Innocent III, references commending bees disappeared from Exultet (the Easter proclamation). The liturgical reform of Vatican II has retained the Exultet Pascal while ratifying the removal of the praise of bees.

CONTENTS

FOREWORD

John Phipps

October 2013

The editor about to collect a tetchy swarm which had set up home in a hawthorn hedge.

All Dressed Up

Beekeeping as a hobby is meant to be a relaxing pastime which can give endless pleasure and enable one to forget everything except for the bees and their intricate and ordered lives. Once the smoker is lit, the beekeeper is already feeling the tensions of his work-a-day life slip away and is thus prepared for a quiet and steady exploration of the hives. No banging of the hives or jerky movements will help to ensure that the examinations - though relatively quick so as not to disturb too much the environmental conditions within the colony - will be completed peacefully, without angering the bees which would almost inevitably lead to unwanted stings.

Well, that is how it is supposed to be. Unfortunately, some hobbyists find that they like to multiply their colonies, so much so that they do not have enough time to enjoy their beekeeping, as the rows of hives get longer and longer. And of course, there is all that dressing up in protective clothing - bee suits, gloves and boots. Too much protection can be extremely uncomfortable in hot weather and also can lead to clumsy beekeeping. - or to examining colonies when the weather is inappropriate. Just like farm tractors with extra wide wheels that can be used on the land no matter what the soil conditions are, the beekeeper, so thoroughly protected can have a go at his bees, whenever, to the detriment of his colonies.

When I lived in the UK, admittedly, I had some foul-tempered colonies. Usually these bees were in out-apiaries and they were ready for me before I switched off the car engine. I had to be fully-protected before I even opened the door and examining the colonies was far from enjoyable. I remember that these stocks had all the bad points which would earn top scores on the negative attributes of a colony on the carefully planned columns of BIBBA hive record cards. The colonies in my garden were not at all bad, though there was always the odd hive which had 'followers' which I ensured was examined last of all.

Two questions, both of which are related - why do so many beekeepers today wear fully protective clothing, and what has made bees so bad-tempered? Whilst the answer to the first question might seem obvious, it may not necessarily be so. So many new beekeepers it seems buy the bee suits, gloves and have Wellingtons ready as a preliminary act to purchasing their bees. How many newcomers to the craft have had the experience of looking at

a colony on a warm, windless sunny day in late spring, wearing only a light veil for protection and with sleeves fully rolled up? To be able to put the palm of their hands on the tops of the frames, to feel the bees crawling over their fingers? I guess very few. A problem definitely associated with bee temper is that the gloves and bee suits are rarely washed between visits to hives and the remains of any venom will be detected by and alarm the bees. The beekeeper naturally then, continues to wear the whole rig believing that the bees are indeed bad-tempered and that full protection is definitely required.

A simple veil/jacket is usually more than enough. The late John Gleed at his apiary in Tain.

The second question has been partly answered: full protection leading to careless manipulations or opening colonies at inappropriate times, plus the traces of the alarm pheromones. Of course there are many other factors, but one of the main ones is the hybridisation of colonies due to decades of importation of foreign queens. Undoubtedly, Buckfast and pure bred Italians are said to be, and can be, beautiful to work with. However, this was not my experience in Italy where the Italian colonies were horrendous - but the bee suits worn by all the beekeepers looked as if they hadn't been washed for a couple of seasons, so their behaviour might have been determined by this.

In countries where very few imports have been able to influence local bee stocks, the colonies are certainly more easily handled - and mostly without a veil. This is certainly the case in Southern Greece - though whilst I always wear a veil, I am often just in shorts, T-shirt and sandals when looking at colonies. There is one exception to this and that is when there is a chilly wind which the bees do not like, so unless I have to, I will not open the hives at such times.

My friend Socratis. He will never wear a veil unless he has been stung on the face a couple of time. Whilst working at my hives, I always wear a veil.

The Editor visiting an apiary in Ukraine - I could put my hands on the top bars and the bees would just walk over my fingers.

A family group - the bees are being transferred from a cork to a modern hive.

In Romania, Bulgaria and the Ukraine, the bees tend to be homogenous, mostly derived from Carpatica stock, and can be easily handled. Bees kept in the home apiary can be handled with impunity, but I have noticed that commercial beekeepers who handle lots of hives and work very quickly (the colonies often being badly jolted when being transported on rough, pot-holed, country roads) tend to cover themselves as much as possible - the temper of the bees being affected by the way in which they are treated.

When one looks at old photos from the past, beekeepers and their families tended not to be wearing veils - I doubt if this was due to bravado or that they didn't want their faces on photos to be obscured in the family album. What has become responsible for the change - the hybridisation of bees, the increase in the production of mono-floral crops, or the way in which beekeepers handle their colonies? Or is it simply that the beekeepers of yester-year were made of sterner stuff?

Mike Thornley, a visitor to my apiary, found the bees in the top bar hive to be very easy to work with.

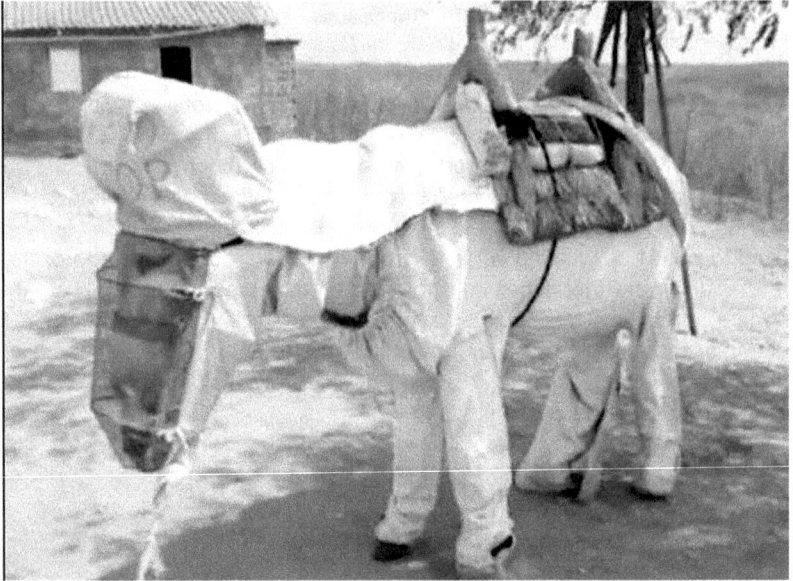
Can you improve on this, Brian Sherriff?

Braula

Looking again to the past, what has become of *Braula coeca*? Has it become extinct in most parts of the UK due mainly to the various treatments used to control varroa. At the time varroa arrived in England Braula was so rife that side by side illustrations of the mite and wingless fly were published together with diagrams showing the feeding positions of the pests on the bodies of bees, the comparisons enabling beekeepers to watch out for and identify varroa. Will Braula resurface again now that many beekeepers are using no chemicals nor organic acids for controlling varroa? An interesting question. Much attention is given to Braula in this edition of the annual - for many new beekeepers it is possibly a creature they have never encountered in their beekeeping.

John Phipps
September 2013

" Went drearily singing the chore-girl small."

Telling the Bees

John Greenleaf Whittier

(1807 - 1892)

a Quaker Poet

Here is the place; right over the hill
Runs the path I took;
You can see the gap in the old wall still,
And the stepping-stones in the shallow brook.
There is the house, with the gate red-barred,
And the poplars tall;
And the barn's brown length, and the cattle-yard,
And the white horns tossing above the wall.
There are the beehives ranged in the sun;
And down by the brink
Of the brook are her poor flowers, weed-o'errun,
Pansy and daffodil, rose and pink.
A year has gone, as the tortoise goes,
Heavy and slow;
And the same rose blows, and the same sun glows,
And the same brook sings of a year ago.
There 's the same sweet clover-smell in the breeze;

And the June sun warm
Tangles his wings of fire in the trees,
Setting, as then, over Fernside farm.
I mind me how with a lover's care
From my Sunday coat
I brushed off the burrs, and smoothed my hair,
And cooled at the brookside my brow and throat.
Since we parted, a month had passed,--
To love, a year;
Down through the beeches I looked at last
On the little red gate and the well-sweep near.
I can see it all now, - the slantwise rain
Of light through the leaves,
The sundown's blaze on her window-pane,
The bloom of her roses under the eaves.
Just the same as a month before,--
The house and the trees,
The barn's brown gable, the vine by the door,--
Nothing changed but the hives of bees.
Before them, under the garden wall,
Forward and back,
Went drearily singing the chore-girl small,
Draping each hive with a shred of black.
Trembling, I listened: the summer sun
Had the chill of snow;
For I knew she was telling the bees of one
Gone on the journey we all must go!
Then I said to myself, "My Mary weeps
For the dead to-day:
Haply her blind old grandsire sleeps
The fret and the pain of his age away."
But her dog whined low; on the doorway sill,
With his cane to his chin,
The old man sat; and the chore-girl still
Sung to the bees stealing out and in.
And the song she was singing ever since
In my ear sounds on:--
"Stay at home, pretty bees, fly not hence!
Mistress Mary is dead and gone!"

John Greenleaf Whittier, 1858

ROT BOTTS
BILL CLARK

●

Are you on bee overload? Is it time to help some other insects?

During my time as Head Warden at Wandlebury Country Park & Nature Reserve near Cambridge, I not only did much of the work, but most of the lecturing, too. Speaking to groups of students in the Cambridge Colleges often ended with a two-way input, which I not only enjoyed but took advantage of. One such talk was to an entomological group, resulting in various visits from the members, when I often found myself being of help with my knowledge of insect habits and their location in the field, so much so, that some professors would ask if I was able to be in attendance before setting a date for their group visits.

One autumn in the 1970s I met entomologist Ivan Perry who was visiting in the hope of finding a rare hoverfly, *Callicera spinolae* feeding on the ivy flowers - since given the common name, Golden Hoverfly. I wanted to know all about it: he had little to impart! It was known to live as a grub in the wet-rot of hollow trees, and had been found in two or three tree species, but was very rare in East Anglia. In 1979 he saw one insect at Wandlebury, and through the eighties and into the nineties saw more, and found grubs, pupae and pupae cases. Most were in the same hollow beech tree, leading him to think they favoured trees in the open, if not full sun. The grubs lived in a layer of precise wetness - not standing water - feeding on the bacteria and microbes for up to four years, and used the upper, drier layer for pupation. My guess was that like many of these grubs, the speed at which they grew to maturity depended on the amount of nutrients available, and besides the correct stage of rotting wood, could be helped or hindered by; decaying fungus, bird droppings, feathers, eggs, bodies or even squirrel urine, to name but a few of the malodorous ingredients.

Unfortunately, during those same years, Wandlebury was hit by gales that took down many of the trees, and droughts that dried out the remaining hollows. I even went to the extent of chain-sawing holes in healthy trees, and pouring pond-water into those and the dried-out hollows. In the spring of 1995 Ivan tentatively brought a plan for my consideration. He had heard of a student, in Scotland, who was filling plastic drink bottles with sawdust and water, and insects were using them! This was marvellous news, I not only had piles of sawdust everywhere, but could now make use of the thrown down plastic bottles. We quickly had some 30 bottles - insulated against frost with plastic bubble wrap - hanging in the trees, and Ivan was soon finding various wriggling grubs inside. Then a spanner was thrown in the works! During his autumn visit, he not only found *Callicera* on the ivy flowers, but saw one laying eggs at the main hollow tree - not inside on the rot as expected, but outside on the bark!

We had to assume that all the egg laying was done this way, and Callicera would probably never choose a plastic bottle. My brain went into overdrive, and by the end of that evening, two of my four dual purpose, 'Mark Two' 20 litre 'Rot Botts' - that had been waiting for Ivan's approval - had been adjusted so that a bark covered beech log was shaped and fixed below the entrance holes, and were hanging up in trees. During the next three years - up to my retirement - I constructed various designs, some to take advantage of available containers, and others to provide habitat for such common species as Drone Flies - *Eristalis tenax*. Before periods of hard frost, I took them down and brought them inside - thus also giving Ivan a chance to investigate them. Although he found good numbers of grubs and pupae, they were nearly all common species - in one hollow tree, the extremely rare, Ethelurgus vulnarata pupae, a predator of other grubs, was found. I will refrain from giving full details of my Rot Bott exhibit at the famous Cambridge Conversazione, when - following the previous evening's staging - the organisers opened up, and were met with a very pungent smell! - Except to say that a couple of professors had an interesting time tracking the source and guessing the composition of the smells and gasses being produced, whilst surmising which one could trigger *Callicera's* interest - evidently from long distance!

Even after I retired I carried on with topping up the water in the 14 containers with my water lance and investigating the many hundreds of empty pupal cases in the five at ground level for another three seasons. As it happens, many of them are the very hoverflies that we often see photographed on flowers and passed off as bees! Although I expected ultra violet radiation to crack up the containers in direct sunlight fairly soon - one white one only lasted five years - eighteen years later there are still a few Rot Botts hanging in the trees; the double purpose ones regularly have birds nesting and there has been a hornet's nest in one. One uncommon hoverfly is associated with them, too - but the old hollow tree has now rotted down to a dry stump, and is only of interest to lesser stag beetles and such.

Ought we to be putting up Rot Botts countrywide?

Rott Botts

Recessed lid keeps out insectivorous birds, holds in moisture, whilst 12 mm holes let in rain water and insects.

25 mm air gap

12 mm water level hole. (Vary up from bottom for different depths of wetness.)

15 to 20 litre buckets. Green or blue plastic weathers best, otherwise paint with dark green gloss paint. Hang from branch or stand on ground accordingly.

25 mm air gap

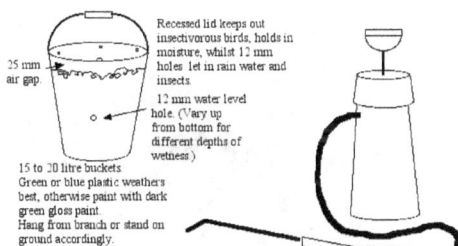

Combined rott bott and nest box. (For those grubs that enjoy other things, besides decaying wood!).

Both handles removed, reconfigured, as one, and re-attached to the lower bucket.

A pump up garden spray - the bigger the better - with the spray nozzle removed, is used to fill the hanging Botts. A length of 10-12 mm plastic tube is attached, extent determined by the highest Bott, and fed through such as - 25 mm domestic waste pipe - 50 or so mm of the tube out of the top end, is then formed into a U with wire, or use a short length of copper tube.

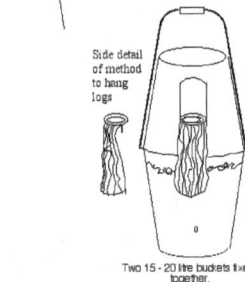

Side detail of method to hang logs

Two 25 litre drums, both tops cut out to leave 25 mm rims, a minimum of four (reasonable diameter) self tapping screws fix them together. Drill the four holes in the (up-turned) upper drum first, and some dexterity is needed to do the fixing through the bird entrance hole! These Botts either had a wire hanger, or were fixed to a back strut, in which case the bottom drum needs be fixed first.

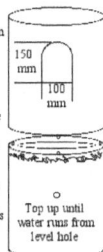

150 mm

100 mm

Top up until water runs from level hole.

Two 15 - 20 litre buckets fixed together.

A D-I-Y guide to making Rot Botts.

MALE (left) and FEMALE (right) DRONE FLIES (*Eristalis tenax*).

(Photo: Wilkimedia Commons, Joaquim Alves Gaspar, Lisboa, Portugal). The insects are good pollinators that particularly favour carrot and fennel flowers. The larvae, because of their looks, are referred to as rat-tailed maggots. They feed on bacteria in any place which is damp and contains putrifying matter. It is believed that these flies were the insects which emerged from the carcase of a lion which gave rise to the saying 'Out of the strong came forth sweetness' (Judges 14:14) - the flies being mistaken for bees. In early times a carcase of an ox was placed in a sealed building in the belief that honeybees might eventually emerge from it.

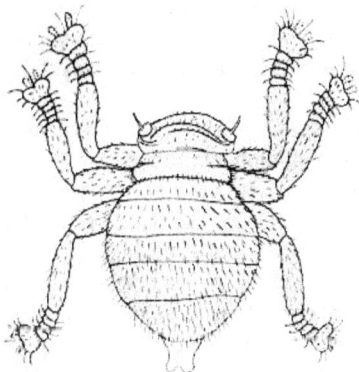

BRAULA COECA - A VANISHING SPECIES?

1. IN HISTORY:

THE BEE-LOUSE, BRAULA COEGA, IN THE UNITED STATES
UNITED STATES DEPARTMENT of AGRICULTURE DEPARTMENT
CIRCULAR 334
Washington, D. C. February, 1925

E. F. PHILLIPS
Senior Apiculturist in Charge of Bee Culture Investigations, Bureau of Entomology

The announcement by the writer (32) of the presence of the so-called bee louse, Braula coeca Nitzsch, in Carroll County, Md, where it has existed for several years, makes it desirable that information be available regarding the relationship of this species to the bee colony. There is nothing on Braula in the American literature except occasional notes on its introduction on imported queen bees with brief statements giving opinions of foreign beekeepers regarding it, usually without reference to investigational work. Even the foreign beekeeping literature usually fails to include the results of scientilic investigations on this species. It has therefore seemed best to summarize the work done, to ascertain to what extent the introduction of this species may be considered worthy of attention, and to list the pertinent literature cited.

The common name bee louse is not an especially appropriate one, since Braula is not a louse, nor does its behavior in feeding suggest even the loose use of that word as a common name. Since the name is well established in many languages, however, there seems no special necessity for protesting its use or of suggesting another common name for the species.

Braula has repeatedly been introduced into the United States on importations of queen bees from foreign countries, and in many cases no effort has been made by the recipients of these queens to remove the parasites before the introduction of the queens. Usually the parasites have disappeared promptly. Since more queen bees have been imported into the United States from Italy than from any other country, Braula is sometimes mentioned in the American literature as the "Italian bee louse." The carelessness of American beekeepers regarding Braula is doubtless due to presumably authentic statements to the effect that the permanent introduction of Braula into the United States is impossible, that it is confined to warm climates, or that it is a quite harmless species. Unless Braula is actually beneficial its introduction is not to be desired, for American beekeepers already have their full share of imported nuisances and pests in the various diseases of the brood and of adult bees, as well as recognized specific enemies of honeybees, all of which are, of course, importations. In addition to the occurrence of Braula in Carroll County, Md, it is also authentically reported to occur in a small area in south central Pennsylvania. Prof A D Whedon, now at the North Dakota Agricultural College, Fargo, N Dak, has kindly furnished the writer with a photomicrograph of Braula taken by him some years ago at Mankato, Minn, from bees located there. The extent and permanence of this occurrence of Braula are unknown. There is no record of importations shortly before these specimens were taken to explain their presence. Beekeepers who find any instances of the occurrence of this species in the United States will confer a favor by sending specimens and a history of the case to the writer.

CONDITIONS IN INFESTED APIARIES IN MARYLAND

The apiaries in Carroll County in which Braula has been found are under the management of successful commercial beekeepers who have watched this species for several years to determine whether or not it is damaging their colonies. They report that it does no damage to strong, healthy colonies in case it is found in them, and in this respect they agree with most European observers. Braula is here usually found on worker bees, rarely more than one to a bee, but under some circumstances they may collect in larger numbers on the queen bees. If, for example, an infested colony of black bees becomes queenless (perhaps in some cases because of the presence of Braula) and if then a young Italian queen is introduced, the insects collect in considerable numbers on the young queen and within a few months she may have the appearance and behavior of an old queen. In the brief examination made of this infestation one drone was found carrying a Braula. Although no thorough examination has been made of all the colonies in the infested apiaries, probably not more than ten per cent of the colonies contain Braula, and it is noteworthy that some of the strongest colonies, and those producing good honey crops, are infested. These apiaries are as well managed as are most commercial apiaries and cannot be classed as badly managed. The results as measured by the honoy crops are good, and it is not the belief of the owners that Braula is reducing the honey crop. There are many poorly kept apiaries in the neighborhood, however, in which the presence of Braula might show a different result.

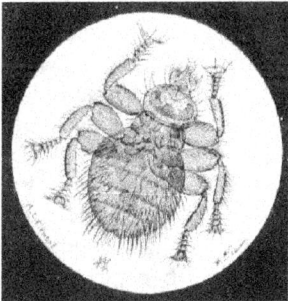

Braula coeca

Feeding on honeybee

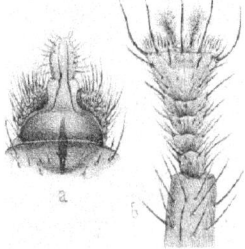

mouth parts and 'claws'.
(from A.-L. Clément, La Nature,
2e semestre 1905, pp. 221-222)

DESCRIPTION OF THE SPECIES

Braula coeca is a wingless, reddish-brown insect having a length of about 1.5 millimeters and a width of about 0.75 millimeter, males being somewhat smaller, on the average, than females. The entire body is covered with numerous stiff, spinelike hairs, at least some of which are said by Massonnat (28) to be connected with nerve endings. These hairs are especially numerous on the head, except on the clypeus and the lower side of the head. The head is flattened from front to rear and is oriented vertically on the thorax, bringing the mouth parts toward the ventral surface of the insect. The antennae have a peculiar structure and are articulated in a deep fossa on each side of the head. Eye rudiments are present, but there are no ocelli. The thorax is discoid and very short on the dorsal surface and is inserted throughout its width on the abdomen. There is no trace of either wings or halters. The legs are of equal length and are long in proportion to the size of the insect. The last tarsal joint of each leg carries a remarkable chitinous comblike structure, divided in the middle with 15 or 16 teeth on each side of the median line, these being modified claws. These combs are serviceable in permitting Braula to attach itself to the branched hairs of its host, which is especially necessary for an animal living on a rapidly moving and flying insect like the honeybee. Each terminal tarsal joint also carries two pear-shaped pulvilli of delicate structure, covered with fine hairs. The abdomen has five visible segments and occupies about 60 per cent of the whole length of the body. It is cylindrical in general shape, tapering to the posterior end, and is flattened less than in most Pupipara. The abdomen of the female when eggs are in formation is terminated by a transparent prolongation in which have been recognized the rudiments of three additional segments, but unless eggs are being formed these segments are invaginated.

Braula is found only on honeybees, although there are statements in the literature that it occurs on bumblebees, doubtless due to incorrect identification of parasitic species thus observed.

CLASSIFICATION OF BRAULA

This insect was first described by Reaumur (33) who briefly discusses the species and its relation to the bee colony. The genus and species were described by Nitzsch (30) who gave it the name *Braula coeca*, classified it with the Diptera on account of the structure of the mouthparts, and allied it with the Pupipara. Various discussions have appeared regarding the exact classification of this species in the series of Diptera of the group Pupipara, and Egger (15) erected for it a special family, Braulidae. Egger corrected certain erroneous statements made by Nitzsch regarding the structure of the antennae and thorax and thereby removed the last doubt as to the alliance of Braula with the Diptera. Miiggenburg {29) shows the relationship of Braula to the Hippoboscidae in the structure of the head vesicle. Until recently the

18

position of Braula with the Pupipara has not been seriously questioned, although, as will be explained later, Braula is not pupiparous. Bezzi (7) was the first taxonomist definitely to remove Braula from the Pupipara and he places it as a subfamily of Phoridae. Dr J M Aldrich is of the opinion that it is better left as a distinct family near the Phoridae.

The synonymy of Braula is not complicated. Costa (13) gave the species the name Entomobis, evidently not knowing of the work of Nitzsch. Bigot (8, vp. 227, 235) suggests that the name of the genus might more appropriately be Melitomyia, as better describing the habit of the species, and later in the same paper the spelling Melitomya occurs, but only the first spelling would stand in the synonym. This latter name is derived from the Attic Greek name, melitta, for the honeybee, as distinguished from the word melissa used by the other Greeks. Since Bigot offers this merely as a proposed synonym for Braula, it need not be discussed further. Fabricius (16) erroneously placed the bee louse in the genus Acarus, based on the figure given by Reaumur.

It is usually stated that there is but one species of Braula, but Arnhart (3), in an effort to explain the diversity of statements regarding its developmental stages, raises the question whether there may not be more than one species. Schmitz (34) has described a new species, *Braula Jcohli*, on African honeybees, but no work has been reported on its development. De Miranda-Ribeiro (27) gives a half tone of the species found in Brazil, in which the head appears to be relatively much narrower than in the European species, and which may be another species; but this, after careful examination, is denied by Lima (22) . Schmitz calls attention to the variation in the number of teeth in the tarsal combs as described and figured by various authors, but since at least some of these illustrations are merely the result of careless drawing it is not well to depend too implicitly on such evidence. The existence of several species of Braula would scarcely be adequate to reconcile the variety of statements which have appeared regarding its development.

DEVELOPMENT OF BRAULA

Until recently it was generally supposed that Braula is pupiparous, although as early as 1858 Leuckart {21) pointed out essential differences between the organs of the female of this species and those of Pupipara, and Miiggenburg {29), a pupil of Leuckart, states that "Professor Leuckart believes, as he has kindly told me, that the eggs of Braula have occasionally been found in the cells of the bee comb." Skaife {35) described the eggs, larvae, and pupae of Braula coeca and for the first time definitely showed that the species is not pupiparous. His conclusions (35, p.8) are:

"Braula coeca is oviparous, not pupiparous, as was hitherto supposed. The eggs are deposited on the brood combs in the hives, hatch out into typical muscid larvae which make their way into cells containing young bee larvae. The larvae feed on food supplied to the brood by the nurse bees, and beyond robbing the bee larvae of a little of their food do no harm. The larvae pupate inside the cells beside the bee pupae; they emerge before the bees do and make their way at once on to the bodies of their hosts. The adults feed on honey, probably supplied to them by their hosts".

An important addition to our knowledge of the breeding behavior of this species was made by Arnhart (2), who shows that development takes place on the under surface of the cappings of honey in the brood combs in special wax tunnels prepared by the Braula larvae - Miiggenburg states that he has never found a larva in the sex organs of the female and further states that the gland tubes which serve to furnish food for the developing larvae of species of Pupipara are lacking in Braula. Skaife confirms Miiggenburg in this point.

The material collected in Carroll County, Md, contained plenty of adult insects, and under cappings of the honey in the brood combs were found eggs, larval skins and pupae clearly identical with those described by Skaife and in the exact position described by Arnhart. The puparium of Braula is not hard, but consists of the last larval skin unthickened; it is not brownish in color as stated by Assmuss (4). Losy (23, 2J) states that copulation of Braula occurs on the queen bee.

GEOGRAPHICAL DISTRIBUTION
Braula occurs in France, Italy, Germany (but not in Hanover and Oldenburg, according to a private communication from Dr. H. v.

Buttel-Reepen), and in the Baltic region, according to Assmuss (4). It is also recorded in Further Pomerania by Timm (5), in Mediterranean countries by Benton (5), on the island of Cyprus by Cook (12), on the authority of Benton, in South Africa by Skaife (So), in Brazilby De Miranda-Ribeiro (7), in the Argentine Republic by Wolffhiigel (5), in Austria by Arnhart (2), in Holland (5), and in Czechoslovakia (Dr. A. Schonfeld in a private communication). It is reported by Assmuss not to occur in northern, middle, or southern Russia, and by Gale (17) as absent from Australia. Cheshire (10) states that it has been introduced into England, but that it rarely survives a winter there. From these records it is evident that the statements which have frequently occurred in American beekeeping literature to the effect that Braula is confined to warm climates are not correct.

FEEDING HABITS
The question of first importance regarding Braula is its exact relation to the bees on which it lives, and the best evidence on this point seems to be the information regarding its method of taking food. Frequent statements have been made to the effect that it takes its food by sucking the blood of the bees on which it lives (5, etc.), and it is frequently mentioned as a parasite. Other writers have referred to it as a commensal of the colony, some of them (S5) stating that it lives on honey. The older writers as a rule considered it

a true parasite. The tongue of Braula has been carefully described by Losy (2S, 24) and by Massonnat (28), who show that the tongue is incapable of piercing the integument of the bee or even of puncturing between the chitinous plates of the abdomen. There are no hard stylets on the proboscis, so that it is evident that in taking food Braula must confine its attention to some source other than the blood of the bees.

The behavior of Braula in feeding has been described by various investigators and the description by Perez (31) is frequently quoted. The original paper has unfortunately not been available to the writer but at least a portion of his results have been translated by Root (31) and are quoted below.

'One day, having captured a bee with one of these lice, I fixed its head with a pair of pincers sufficiently to keep it unmovable, and to capture the small parasite easily. Both it and the bee were left for a while on the table in my studio, under glass. When I returned to them I was not a little puzzled to see the parasite in the most vivacious and strange agitation. Seated on the fore part of the bee's head it was moving about with incredible vivacity, as though possessed of veritable rage. Now it would go to the margin of the bee's cap, with its fore feet raised, stamp and scratch as hard as its weakness would allow at the base of the bee's lip; then it would suddenly run back to the insertion of the antennae to renew its impetuous attack immediately. I was quite taken up by my first surprise, when I suddenly saw all this fury turned to perfect calmness, and the little animal squatted on the edge of the cap and bent down its head to the bee's mouth, which was slightly trembling, and sucked up a drop of moisture. I instantly understood. The movements I had just witnessed were preparatory to the animal's meals. When the louse wishes to feed it goes to the bee's mouth, where the motions of its feet, armed with bent claws, produces a tickling sensation, perhaps disagreeable to its host, but at least provoking some movement of the buccal organs, which slightly open and release a small drop of honey which the louse at once licks up.'

Thus the *Braula coeca* is not a real parasite of the bee in the true sense of the word. It is rather a guest — queer, if you like thus to consider it, like so many others existing among animals.

Losy (23, 2Jj) has studied the mouthparts of Braula in detail as well as the feeding habits. His two papers are in Hungarian, but a good review by Gorka (2^) gives what appears to be the essential part of his results. Since this review is important, the main parts are here freely translated:

"The parasites are mostly on the queen and first go over to the workers when they undertake the feeding of the larvae, when Braula nibbles at the food which is conveyed to the brood. As soon as this feeding is ended they are all found on the queen, on which mating also occurs. Their number by the end of November becomes so great that the queen is in danger and she becomes weakened and in late fall she perishes. In greater degrees of infestation (in unclean colonies) this may occur even in summer, and may result in the death of entire colonies. The mouthparts of Braula form a sucking organ which is adapted to the mouthparts of the bee in an astonishing manner. The bee louse sucks its nourishment from the outstretched mouthparts of the bee. It perceives the stir which occurs from the movement of the chitinous parts of the skeleton during sucking, which tells it that it can again enter the mouthparts of the bee. Then it quickly runs over the back of the mouthparts as the bee holds out its tongue for the sucking of food, sucks up and swallows the food, or if food is allowed to drop into the cells of brood Braula takes the sweet food arising from the glands of the bee. The bee louse remains standing on the open jaws and the upper lip, and when the jaws of the bee are about to be separated its sucking organ is opened. This separation is assisted by the Braula so that it wedges with brushlike bristles in between the mouthparts of the bee and separates them, then it stretches out its proboscis and reaches it to the back upper surface of the tongue. As soon as the tongue of the bee is in motion, the horizontally held beak of the Braula proboscis reaches into the cavity at the base of the bee tongue which is then brought forward. Here it is pinched under the paraglossa, is broadened, its bristles are ruffled up, and with the two supporting bristles of the lower lip it spreads the paraglossa of the bee apart and in this way it not only prevents the drawing back of the tongue, but it also holds the basal part of the tongue open. This occurs for the reason that beneath the base of the tongue the external opening of the canal from the salivary gland is found. Through this gland opening, because of the irritation of the unusual penetrating body and the unusual saliva, saliva is poured forth reflexively, which the Braula sucks up. Braula is therefore a parasite which has become adapted to the organism of the queenbee and is a burden and torture of the first order to them."

In the examination of Braula from Maryland, smears were made of the contents of the alimentary canal, and in one individual 11 peculiarly shaped pollen grains were found, while in the alimentary canals of the bees from which the Braula were taken pollen from the same plant source (unidentified) were abundant. Whether pollen is an important constituent of the diet of Braula is not known and it may have entered merely in association with liquid food taken in the way described by Losy. The liquid portion of the diet could not be identified from the smears.

It seems to be almost the unanimous opinion of beekeepers of European countries that Braula occurs in weak colonies and especially in those not properly housed and cared for. It is also stated (20) that it is more abundant in poor seasons. Emphasis is always placed on the necessity of keeping the hives, and especially the bottom boards, clean, and obviously these precautions are not taken except by the best beekeepers. The question arises, therefore, whether the weakness of colonies containing Braula is the cause or result of the infestation. If we look on Braula merely as an undesirable commensal, its damage would probably be slight, although the bees usually seem to make little effort to remove it. If it is actually, as claimed by Losy, a parasite of highly specialized habit, then it may be a dangerous parasite to the welfare of the colony. Since in the one case where this insect has been observed in the United States it does not seem to be causing much damage, the extreme view of Losy would not seem to be supported,

An interesting discussion of Braula in the infested areas is given in a private conimunication (January 12, 1924) from Dr. Ludwig- Arnhart of the Osterreichische Imkerschule, Vienna, who says in part:

"As far as my own experience and knowledge go, Braula becomes harmful only if found in large numbers in the hive. A single one will not injure the queen, the worker, or the drone. The last is attaokod the least. They are most frequently found on queens, which are weakened to a great extent by large numbers of these lice and perish easily during the winter season".

NUMBER OF BRAULA FOUND ON ONE BEE

It appears that usually there is not more than one Braula on each worker bee, although there may be more on the queen bee, if, as described by Losy, there is a migration from the workers to the queen at the close of brood rearing. This might partially account for the wide discrepancy in the reports on this subject. Assmuss (4) states that in his experience there is usually only one to a bee, but that they may occur in much greater numbers. They occur on workers, drones, and especially on queens. Hammer (18) reports taking 187 Braula from a queen and at a later date 64 from the same queen. Cheshire (10) reports removing 6 from a queen in England; Cook (12) reports that Benton has taken as many as 10 from a single bee; Kramer and Theiler (20) report that as many as 60 have been taken from a single queen; Benton (5) reports having removed as many as 75 from a queen at one time ''although ordinarily the numbers do not exceed a dozen"; while Marboud (26) reports that he took off 31, the next day 33, two days later 43, and continued until he had removed a total of 371.

Tinun (36) questions the accuracy of these extravagant statements. The bee louse is relatively large in proportion to the size of the honeybee and it would seem impossible for a queen to carry such great numbers as those reported. He expresses the belief that these observers have mistaken either mites or triungulin larvae of Meloe for Braula, although there would seem to be little excuse for such an error. It is certainly the case that Braula usually occurs singly on worker bees and that if larger numbers occur it is almost exclusively on queens. In Maryland two Braula have not been observed on a single worker and they are rare on the queens, except under the conditions previously described.

PERCEPTION OF LIGHT BY BRAULA

The specific name *coeca* was given this species by Nitzsch on the assumption that it is blind. Miiggenburg (29) points out that it is not blind, for despite the previous statements, it has two small eyes which lie above the antennae. From their situation these eyes represent the compound eyes of the Diptera. Their dioptic apparatus is only very slightly developed. The chitin of the head covering over the place in question is thinner and transparent . . . and shows no trace of facet formation. Ommatidia are not found under the imperfect cornea. From sections we perceive, as in early developmental stages of insect eyes, a mass of hypodermal cells which show a tendency to radial arrangement. Pigment is not present. However, a thin optic nerve extends to this rudimentary eye from the supra-oesophageal ganglion which increases near its connection with the same to a small ganglion.

Massonnat (28) describes this eye structure in still more detail and figures the various internal parts, but, unlike Miiggenburg, he claims to find traces of pigment. Timm (36) also describes the presence of eyes. Von Buttel-Reepen (37) pointed out to beekeepers the incorrectness of the statements that Braula is blind, but this same error has since been repeated in more recent beekeeping papers.

No physiological work has been reported to determine whether Braula actually responds to light stimuli.

REMEDIES

It is commonly stated in European beekeeping literature that Braula occurs in weak and badly managed colonies, especially in those the hives of which are not kept clean. If these statements are correct, preventive measures would seem to be of far greater importance than methods for the removal of Braula. Various methods are given for its removal. An early method was to pick them from the queens by means of a feather, some suggesting first dipping it in honey to cause them to adhere well. Arnhart (2) uses small pointed sticks dipped in honey in the same manner. It is also recommended (6) that the queen be removed from the colony and gently smoked with tobacco smoke, which causes the stupefied Braula to drop off, after which they may be destroyed, and since she will probably collect more when returned to the hive, it is usually recommended that this be repeated at intervals. This method has its faults, since the smoking of the queen may cause the bees to ball her when returned. The placing of a small bag of napthalene on the bottom board of the hive is said (11, 19, 25) to cause the insects to drop from the bees to the bottom board, from which they should then be removed before any of them recover. Smoking with saltpeter or Lycopodium, old methods for stupefying bees, have been recommended (9), as well as oil of turpentine (19) on a cloth on the bottom board, carbolic acid (V^) similarly used, and incense powder (11), these all being used because Braula apparently succumbs to such fumigation before the bees do. Zander (39) , however, points out that naphthalene may not only dislodge the Braula but may also drive the bees from the hive. Since there may be developmental stages not affected by the first treatment, the operation should, after an interval, be repeated to remove those which in the meantime have reached the adult stage.

A safe method has just been recommended (1) which if satisfactory has more in its favor than the methods just suggested. This

is that, since weak, listless colonies are the ones which harbor Braula, the infested colony should be opened in the evening and sprinkled thoroughly with honey water. Then, in cleaning up the dilute honey, the bees are said to remove the Braula. None of these methods has as yet been tested in the United States.

CONCLUSIONS

It appears from the available information that Braula is not a serious pest of the apiary and that no great harm is to be anticipated from its unfortunate establishment in this country. It is, of course, not certain that it will remain here. This introduction, however, should not be considered as unimportant, and wherever Braula occurs steps should be taken to eradicate it from the infested colonies. With such a visitant of the bee colony, it is impossible to determine in advance what effect it may have in some other portion of the country, and every means should be taken to eradicate it if possible, especially since the infested area seems to be small at present. Inspectors of apiaries who find this insect in colonies inspected by them will do well to recomimend its removal. More work is needed on the life history and especially on the feeding habits of the species.

LITERATURE CITED

(1) Anonymous. 1923. [Rundschau] Bienenlause vertreiben. Schweiz. Bienen-Zeitung, n. f., vol. 46, no. 11, p. 530.

(2) Aenhart, Ludwig. 1923a. Die Larve der Bienenlaus in den Wachsdeckeln der Honigzellen. Bienen-Vater, vol. 55, no. 6, pp. 136-137, 1 fig.

(3) 19236. Zur Entwicklungsgeschichte der Braula coeca Nitzsch. Zool. Anz., vol. 56, no. 9-10, pp. 193-197, 1 fig.

(4) Assmuss, Eduard. 1865. Die Parasiten der Honigbiene und die durch dieselben bedingten Krankheiten dieses Insects nach eigenen Erfahrungen und dem neuesten Standpunkt der Wissenschaft, 56 pp., 3 col. pis., 26 figs. Berlin.

(5) Benton, Frank. 1899. The honeybee: A manual of instruction in apiculture, Bui. 1, Div. Ent., U. S. Dept. Agr., 118 pp., illus.

(6) Bertrand, Ed. [1904]. Conduite du rucher. 9me ed., 288 pp., illus.

(7) Bezzi. 1916. Riduzione e scomparsa delle ali negli insetti ditteri. Rivista di Scienze Naturali, vol. 7, pp. 85-182.

(8) Bigot, J. M. F. 1885. Dipteres nouveaux ou peu connus, XXXV. Ann. Soc. Ent. de France (6), vol. 5.

(9) Boise, P. 1890. [Note sur Braula caeca {coeca) Diptera:] Communication a la Soc. Ent. de France. Bui. des Stances et Bui. Bibliograph, de la Soc. Ent. de France, (6) vol. 10, pp. cc-cci.

(10) Cheshire, Frank R. 1888. Bees and beekeeping. 2 vols. (See v. 2, pp. 577-578.)

(11) Clement, A. L. 1905. Le Braula coeca. La Nature, vol. 33, no. 1684, pp. 221-222, 4 figs., September 2.

(12) Cook, A. J. 1888. The bee-keeper's guide or manual of the apiary. 461 pp., Lansing, Mich,

(13) Costa. 1845. Atti del. Reale Instit. Incorrag., vol. 7.-

(14) Cowan, T. W. 1911. The British bee-keeper's guide book. 20th ed., 75th thousand, 226 pp., London.

(15) Egger, J. 1853. Beitrage zur bessern Kenntniss der Braula coeca Nitzsch. Verhandl. des zool. — -bot. Ver. zu Wien, vol. 3, pp. 401-408,

(16) Fabricius. 1794. Ent. Syst., vol. 4, p. 432, no. 37. 1905. Enemies of bees. Agr. Gaz., N. S. W., vol. 16, no. 5, pp. 489-492.

(18) Hammer, G. 1858. Die Lausesucht unter den Koniginnen. Eiehstadt Bienenzeit., vol. 14, no. 1, pp. 10-11.

(19) HOMMELL, R[OBERT]. 1919. Apiculture. 501 pp. Paris.

(20) Kramer, U., and Theiler, J. 1910. Der schweizerische Bienenvater, 7te Auf., 322 pp. Aarau.

(21) Leuckart, Rud. 1858. Die Fortpflanzung und Entwicklung der Pupiparen nach Beo- bachtungen an Melophagus ovinus. Abhandl. der naturforsch. Gesell. zu Halle, vol. 4, pp. 145-226, 3 pis.

(22) Lima, Costa. 1923. Nota sobre a "Braula coeca" Nitzsch no Brasil. Revista Brasileira de Apicultura, vol. 2, no. 4-5, pp. 51-55.

(23) LosY, J6sef. 1902. A meh es mehtetti egyttelese. Rovartani lapok havi folyoirat kiilonos tekintettel a hasznos es Kdrtekony rovarokra, vol. 9, pp. 153-156, 175-180, 5 text figs.

(24) 1902. [Same title.] Kiserletiigyi Koz-lemenyek, vol. 5, pp. 163-204 6 figs., 3 pi. Budapest. Review bv Gorka, Zool. Centrabbl., vol. 10, (1903), pp. 840-842.

(25) Ludwig, August. 1906. Unsere Bienen. 831 pp., 4 Kapitel. Die Bienenfeinde im iibrigen Tierreich. Berlin.

(26) Marboud. 1907. Le pou des abeilles. [A Revue agricole de l'Ain.] L'Apiculteur, vol. 51, pp. 342-344.

(27) Miranda Ribeiro, Alipio de [Secretario do Museu]. 1905. Braula coeca Nietsch. Archivos do Museu Nac. do Rio de Janeiro, vol. 13, pp. 155-161, illus.

(28) Massonnat, Smile. 1909. Contribution a l'etude des Pupipares. Ann. de l'Univ. de Lyon, n. s., vol. 1, Sciences, medicine, fasc. 28, pp. 1-388, 112 figs. 7 pis.

(29) MTJGGENBURG, FRIEDRICH HANS. 1892. Der Rtissel der Diptera pupipara. Archiv fiir Naturgeschichte, vol. 58, no. 1, p. 287-332, pis. 15, 16.

(30) Nitzsch, Chr. L. 1818. Die Familien und Gattungen der Thierinsekten (insecta

epizoica); als Prodromus einer Naturgeschichte derselben. Magazin der Entomologie herausgegeben von E. F. Germar und J. L. T. F. Zincken (Sommer), vol. 3, no. 9, pp. 261-316.

(31) Perez, J. 1882. Notes d 'apiculture. Bull. Soc. d'apic. de la Gironde. Bordeaux. Trans, in part: Root, ABC and XYZ of Bee Culture, 1920 ed.

(32) [Phillips, E. F.] 1923. [Note on occurrence of bee louse in Maryland.] Monthly News Letter of Bur. Ent., no. 113, September, 1923. Copied in Notes on Apiculture, Jour. Econ. Ent., vol. 16, no. 6, p. 562; Insect pest survey bulletin, Bur. Ent., vol. 3, no. 7 (October) , p. 304.

(33) Reaumur, 1740. Memoires pour servir a l'histoire des Insectes, vol. 5. Paris.

(34) SCHMITZ, H. 1914. Eine auf der afrikanischen Honigbiene schmarotzende neue Braula- Art. Archiv de Zool. expert, et Gen., vol. 54. Notes et revue, no. 5, pp. 121-123.

(35) Skaife, S. H. 1921. On Br aula coeca Nitzsch, a dipterous parasite of the honevbee. Trans. Rov. Soc S Africa, vol. 10, pt. 1, pp. 41-48, 11 figs.

(36) TiMM, Paul. 1917. Zur Lebensweise der Bienenlaus {Braula coejca, Nitzsch). 39, Bericht des westpreuss. bot.-zool. Ver., pp. 1-5.

(37) V. Buttel-Reepen, H. 1919. Einiges liber Bienenschadlinge imd die Bienenlaus, Braula coeca, Nitzsch. Bienenwirtsch. Ztrbl., 13, 14, 21/22.

(38) WOLFFHTJGEL, KURT [Dr.]. 1910. Braula coeca Nitzsch en la Republica Argentina, Anales de la sociedad cientifica Argentina, vol. 69, no. 3, p. 124.

(39) Zander, Enoch. 1921. Krankheiten und Schadlinge der erwachsenen Bienen. 60 pp. Stuttgart.

(The price of this paper was 5 cents in 1925. Ed)

2. STATUS OF BRAULA WORLDWIDE

Although Braula is primarily an inhabitant of countries with a Mediterranean climate, its range has been far and wide over the last century. Whilst cold winters can kill Braula, the eggs and pupae can survive within the warmth of the colony's winter nest. Braula has been found in Brazil, the USA, England, Scotland, France, Spain, Italy, Russia, Tasmania, Bulgaria, South Africa, Mauritius, and Sicily as well as other countries but is absent from Australia, New Zealand and Papua New Guinea. In Britain Braula was brought to the attention of beekeepers by Herrod-Hempsall, Wedmore and Manley.

Braula is considered to be more of a nuisance than a threat, but in Australia if found in hives its presence has to be reported within twelve hours.

3. BRAULA IN THE ORKNEY ISLANDS
Sue Spence, September 2013

Many of us beekeepers in Orkney have braula in our hives, spread about through swapping brood combs of stores between colonies and apiaries. It doesn't seem to be a huge problem. Young, newly-started queens can sometimes be seen carrying a huge burden but this does not seem to last very long nor to inhibit their laying. I don't know why; whether it's as a result of new bees hatching thus increasing nurse and house bees, or because by the time a young queen is up and going here the weather is getting cooler and the mites are slowing down.

I have tried various treatments over the years; the old books recommend that you smoke a pipe of tobacco and whilst examining your bees you gently blow smoke from your pipe over the infected bee - apparently the smoke is an irritant and the wingless flies abandon the host falling onto the floor (This is all very well if you can smoke a pipe whilst wearing a veil - or even manage to smoke a pipe at all without choking to death!). A little pipe tobacco in the smoker used sparingly is quite good; you have to slide a piece of stiffish paper into the hive entrance to cover the floor to catch the stunned mites and remove the paper before they all hop back onto their hosts.

I know of one beekeeper who tried a full tobacco smoke treatment on a hive, blocking the entrance and leaving the colony overnight, the result was an unqualified success in eradicating the braula - but all the bees died too.

Crystals of menthol are effective if the weather is warm enough for the vapour to penetrate the hive, this has to be done after the honey crop has been taken, so it is not often an appropriate treatment here in Orkney where our average summer temperature is 12 degrees (our maximum ever was 26). It's done in much the same way as the tobacco smoke but with the crystals being placed on to the top bar of the brood frames.

All three stages of Braula can die over the winter as stored honey is consumed (I believe the eggs are laid beneath the capping), and the weather gets colder. I suppose if a colony were to consume all its winter stores the braula would die out in that colony, but obviously this is not a good idea.

Renewing brood frames and throwing away very old comb whilst a good practice does little to stop braula. Re-hiving, with new foundation in the early spring is about the only way of getting rid of the infestation (for a while) and good hygiene can help, too. I don't know if it moves on to flying bees, particularly drones which are welcomed into all hives and thus help to spread the pest.

Simon Croson

STUNG BY A NEW CAREER!

BY SIMON CROSON
LINCOLNSHIRE BKA –
FORMER CHAIRMAN, BBKA BASIC
ASSESSOR, NATIONAL HONEY SHOW
LECTURE CONVENOR

It's amazing how things change and more importantly how things change us. In 2006 I took up beekeeping and in 2007 wrote an article for the BKA about being "Stung by a new Hobby". Well, time has moved on and my Beekeeping journey has taken so many roads, allowed me to meet so many great people and given me the chance to stop off at some amazing places and allowed me so many opportunities that I never expected in my wildest dreams. Beekeeping, it's more than a Hobby!

After that initial Beekeeping course my appetite was very much stimulated and I knew that there was such a world of possibilities I just had to start the wheels turning. My first active season saw me drawn in to the show arena at County and National level with very rewarding results. I will never forget my first Card for liquid honey that was just shy of the correct fill level, a point put over by the judge in a very encouraging way that suggested that the honey was good enough for a top three card. My photographs won immediate recognition and it was this that has propelled me on my beekeeping journey. Winning a 1st place at County level gets people looking over their glasses at you – a 1st at The National brings on a smile and handshake and a Gold Medal at Apimondia – well, we are back to that over the glasses look! I initially took the photographs for personal research as I could review that which I had seen at the hive in the relaxed conditions of home and learn so much – my first good photo showed varroa on a worker. As I began to share the photographs the reactions were very satisfying and soon led to many being published and many winning further awards at County and National but the highlight to date has been the success at Apimondia 2011 in Argentina where 4 photographs returned the Gold Medal and a Diploma in Photography Innovation. As I write, the entries for Apimondia 2013 have just left.

Alongside the photography was my beekeeping progression from 2 to 20 hives within 2 years and also progression within the BBKA examinations arena which has been at sometimes very challenging but equally as rewarding as the practical beekeeping. Anyone who has yet to start off along this road, please do, as your beekeeping will be lit up by the vast amount of enjoyable study required. Having now achieved the BBKA Intermediate Certificate, General Husbandry and becoming a BBKA Basic Assessor I can see the BBKA Master Beekeeper in my sights but the road is longer and harder, and so it should be !!

With my old occupation(RAF Engineer) leading me into professional instructor duties and professional qualification(DTLLS) I knew that I wanted to turn this asset into a Beekeeping benefit and began teaching beekeeping and also delivering a number of evening talks on our craft.This has grown significantly with many more possibilities in the pipeline. Meanwhile in the background the honey show awards kept rolling in, although I am still in envy of the highest awards at The National Honey Show. Involvement in The National Honey Show exec committee allows me the chance to convene the lecture programme. My 1st for photography and wax is a step in the right direction; no wonder they call it the best honey show in the World. The number of hives began to grow rapidly and a hobby soon began to take hold and in 2010 it was looking like beekeeping was going to be a full time occupation. Winning a BBC Good Food Award for the honey products was evidence enough that my honey and the product was worth a gamble to go full time so the path was a lot clearer- we established The Artisan Honey Company.

The photography was still providing opportunities and when you have some of the biggest American Beekeeping Authors paying the highest of compliments, you have to listen. An invite to EAS 2012, in Vermont, was another catalyst and in front of a room of upwards of 150 – I was in my element delivering a talk on Beekeeping Photography but conscious of the audience – the plaudits afterwards were so kind and a return to America, Florida Bee College this year consolidated my faith. My talk there was in full glare of Jamie Ellis of the University of Florida Bee College Senior and acclaimed author, who offered his congratulations.

As a professional beekeeper there can be times when the bees are not so demanding – I haven't found that time yet – and additional undertakings can be very welcome. The trip to Florida was combined with a Sue Cobey Instrumental Insemination course, which will prove invaluable in the coming years as I develop my encouraging queen rearing programme in conjunction with some very supportive peers.

The teaching is still going strong and many other opportunities are coming into sight, the next big ones are Honey Hunting in Nepal when I get to mix my two loves of photography and beekeeping by leading a trip by Bees Abroad into the Western Himalaya to witness the honey harvest; by the time this is being read we will be preparing to leave, I hope to have features in a number of media outlets, another return trip to South Florida in summer 2014 will see the talks on photography evolve into practical workshops in the apiary.

The hive numbers have risen to almost 200 and I am pleased to say that the 2013 season ended with the majority of my colonies being headed by my own raised queens. Along with honey, wax, teaching, photography, queens and nucs also provide a welcome and varied income.

The journey from Hobby to Professional, which is still moving uphill, has been one of opportunity and amazement but one that I would recommend to anyone – whether it's beekeeping or something else that lights your candle! Grasp the opportunity and make it your own.

Simon Croson's website:
http://www.theartisanhoneycompany.co.uk/gallery.php

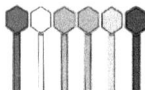

FROM LONG JOHN SILVER
TO A U-BOAT CAPTAIN

OLD IS OFTEN BEST

John Kinross

Once again it is time to call together the Taranov Board to their annual meeting. For new readers it means 'Tomes and Reprints Allowed (Not on Varroa)' and the three meet in the local pub to discuss the current year's collection. Professor Dripitoff is getting elderly and has given up driving, so I have to fetch him. Ina Strainer is there already in her battered Morris Traveller full of ekes, a spinning wheel and countless bits of hives. She even had a swarm once in the back of her car whilst driving down the M50 towards Wales. Luckily, the car behind was driven by a beekeeper who helped her to catch it and together they shared the honey out later - amongst other things, I gather.

Firstly in my list is *Ken Steven's* vast tome, 763 pages, **Alphabetical Guide for Beekeepers** (NBB £30) which looks so much

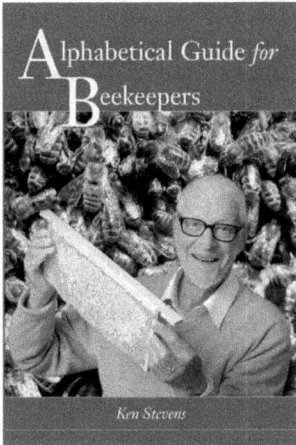

Alphabetical Guide *for* Beekeepers

Ken Stevens

better now than it did originally when the author tried to produce it on his own. It is a nice to read that BBNO has an entry - but at their old address, and that a bee has three eyes arranged in a triangle. This would be useful in humans, but more especially for one-eyed characters like Silver, L J, or Nelson, Lord H as they would then have a spare. Our U-Boat commander might have found it useful, too, as well as British submarine commanders. It must have been very tiring looking through that long periscope.

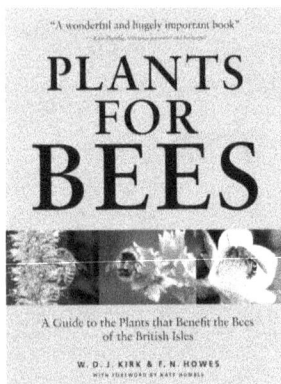

"A wonderful and hugely important book"

PLANTS FOR BEES

A Guide to the Plants that Benefit the Bees of the British Isles

W. D. J. KIRK & F. N. HOWES
WITH FOREWORD BY KATE HOWES

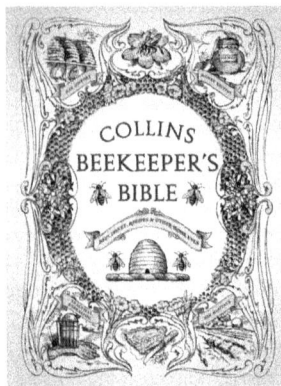

COLLINS BEEKEEPER'S BIBLE

Kirk and Howes (the late Mr F N) have produced **Plants for Bees** (IBRA £25) - a large, heavy book based on *Howes'* original **Plants and Beekeeping** which came out in hardback with B&W illustrations, then in paperback with none at all. Now there are excellent colour plates in this book, taken by a variety of people including Chris O'Toole (whose bumblebee poster is still available from BBNO), Ings, Kirk, Owens and Early. The Index is excellent though the Glossary is a bit small - but it is nice to know that an anther is not a horn of a drunken deer.

Ina often gets asked for suitable books for presents, especially for a family of beekeepers. She likes the Collins' **Beekeeper's Bible** (Collins £30) as it has some wonderful recipes including one for Baklava - which she said was what the people's leaders at Pompei said when they were to be covered with lava. The Professor said he always wore one when riding his Harley. It is a mystery though why the publishers of this nicely-produced book don't give an inkling as to who wrote it, but at the end Nicola Charlton is mentioned as copy editor and Jenny Heller as editorial director, so it must have been a team effort.

Three New Outstanding Books

The last few months of 2012 has seen the publication of three new beekeeping books which will help the student of bee anatomy. Firstly, *Dr Ian Stell* has written and published his own book **Understanding Bee Anatomy: a full colour guide** (Catford Books, £28). It has 200 pages, closely packed with a lot of intense anatomical information. The Reader needs a Biology 'A' Level or degree to make the most of its contents and

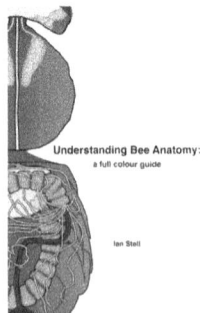

Understanding Bee Anatomy:
a full colour guide

Ian Stell

those of us, like me, who gave up science in the 4th form will need a Glossary. However, one of my readers who worked for six years in a slaughter-house was quite delighted with it.

Secondly, and more straightforward, *Bob Maurer* has written **Practical Microscopy for Beekeepers** which is published by Bee Craft for £15. Maurer recommends two microscopes for beekeepers, a compound with one eye-piece (for the Long John Silvers) and a dissecting one with two eye-pieces (for ex U-Boat captains?). It doesn't help if you have specs as they mist up as soon as you look into a microscope, but maybe it is only me that has this problem. I am sure that those of you with steady hands and good eyesight will find this guide invaluable.

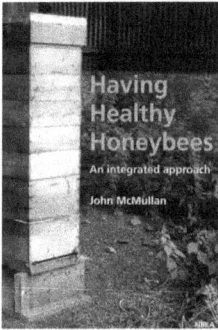

Thirdly, from Southern Ireland, *John Mc Cullen* has written **Having Healthy Honeybees** (FIBA, Eire, £16) which is a neat paperback with several colour plates and a Bibliography of ten pages but alas no index. He is excellent on bee diseases but no wiser on CCD than other writers. There is a useful section on National Disease Policy - but more on what is notifiable than how to cure varroa, for instance.

Please Mr McCullen when you reprint your book, cut down the biography and add an index and glossary. All beginners need them.

Two reprints from NBB have just appeared. *Richard Taylor's* **The Comb Honey Book** (NBB £11.95) which first appeared in the 1970s but, with Killion's book on the same subject out of print, will be an invaluable guide for those attempting to produce comb honey. The other, **Producing Royal Jelly** by *R P van Toor* (NBB £12) - the task all beekeepers know is a very specialist job; the author, who worked for the New Zealand MAFF in 1986-90, knows more about it than most. Once again, there is no index, and some of the illustrations are very black, but there is a useful and detailed 'Contents'.

A WINTER PROJECT FOR THE WORKSHOP

A BEEKEEPER'S BOX

JOHN PHIPPS

A long, long time ago, I can remember the late George Hawthorne, the one time CBI for Berkshire and collector of all sorts of items which he described in his 'Collector's Corner' column of The Beekeepers Quarterly, bringing a box to a meeting in which he kept his everyday beekeeping paraphernalia and revealing its contents for all present. What surprised me was just how many items could be collected together and arranged so that they were easily found for a beekeeping task.

Donald Sims in his book 'Sixty Years with Bees' (page 123) has a photograph of his box, the interior of which he had made into several compartments to house the various pieces of equipment.

For the last forty years I have been meaning to make such a box and this summer, just before the temperature outside reached an unbearable 35C, I set to work on my project.

1. The Box:

This is made from a Langstroth Brood Body and has been converted into sections so most of the bits and pieces I might need for an apiary visit will be immediately to hand.

The box has been modified by slicing off a section with a circular saw. It is big enough to hold two frames of live bees (perhaps with a queen or queen cells for transferring to another apiary) or for removing a couple of frames of freshly sealed honey for pressing. This section of the box was given a new side wall of thin exterior ply and a similar piece was attached to the box it was cut away from. Both the main box and the section have separate exterior ply floors. It is important that the section for transporting bees is detachable - otherwise it would be difficult to empty it of bees.

Importantly, the smoker section is surrounded by a metal shield - both around and on the floor.

The handle, which can be moved from one side to the other, is made from a broom shaft.

2. The Contents

In/on the box - veil, smoker, rotten hessian sacking for smoker fuel (-though often I carry a bag of pine needles and pick some fresh rosemary or sage both to cool the smoke and to give it a good aroma).

Left - Right:

Tray containing queen cages, queen cell protectors, matches and coloured electrical insulation tape (a piece of the coloured tape can be stuck onto a hive/nucleus to denote the age of the queen using the international colouring system).

Against the box - Honeybee Disease Recognition Cards - just published (June 2013) by Norman Carreck of the International Bee Research Association - the full colour cards give you immediate access to the symptoms of honeybee pests and diseases so that checks can easily be made in the field.
American Foulbrood Diagnostic Kit (Vita Europe).

Tape for sealing any gaps when moving bees.

Lighter for smoker.

Drawing pins.

Leatherman tool kit.

Apipen - adrenalin syringe in case of a life-threatening allergic reaction.

Various hive tools - which can be used for levering apart hives, scraping wax from frames, lifting frames from hive rebates, etc. I prefer the J-shaped hive tool as Langstroth frames have short lugs. The hive tool on the far right is from the Ukraine. It has one feature which I like - the part of the hive tool which is gripped has a covering of plastic which means the sharp, squared-off edges don't cut into your hand. Whilst this might not bother beekeepers who wear gloves, after going through many hives without gloves the hard edges really make themselves felt. Putting insulation tape round standard hive tools might solve the problem. Unfortunately, the Ukrainian hive tool isn't well-tempered and bends if too much strain is applied.

Waterproof notebook and pencil (yes, not only can you write in the rain, but also completely underwater!).

Magnifying glass.

Finely-ground sugar for checking for the presence of varroa mites.

Bee brush.

Water - for washing sticky hands - or dowsing fire in smoker.

Sponge from which small pieces can be removed to plug any small gaps in hives.

Additionally:

A First Aid Kit should be added containing a good pair of scissors and tweezers (in Russian apiaries it is mandatory to have a first aid kit).
I haven't made a cover for the box as, firstly, it isn't necessary as the box is usually kept under cover or in the car (fortunately, it fits perfectly in the rear of our Fiat Panda), and secondly, I didn't wish to add to the weight of the box. Before leaving the apiary the fire in the smoker is dowsed with water.

THE MELISSA GARDEN

BY BARBARA AND JACQUES SCHLUMBERGER AND MICHAEL THIELE

The Melissa Garden is a honeybee, native pollinator (there are 1700 species of native bees in California) and habitat garden sanctuary in Healdsburg, California, at the western edge of the Russian River Valley, on top of a ridge at 850 feet in elevation. Four gardens planted with many exuberant and profuse flowers for nectar and pollen forage are situated in the center of a pristine 40-acre ranch that is lush with native vegetation. The Melissa Garden is a new project that began in the fall of 2007 at their home; the garden initially began with a concern about the plight of honeybees, but when Barbara and Jacques realized that many native bees, butterflies and bird populations are also declining, they wanted to embrace their needs and the necessity to educate people about their status as well. The goal is to provide honeybees, native bees and other pollinators with an almost year-round source of floral resources- free from pesticides. Studies have found that native bees and honeybees both benefit from feeding on a variety of flowers, so almost year-round the garden is kept filled with an abundance of annuals, perennials and shrubs that offer attractive pollen and nectar to insect visitors. The garden is composed of a mixture of plants native to California, many Mediterranean plants and others that are appropriate for the site and climate.

New (old) hive designs

Once we approach beekeeping in the context of the "Bien", which represents the undividable entity of the hive, our methods and hives will change accordingly. There are currently three alternative hive designs (and more to come) used at "The Melissa Garden"

One-Room-Hive (Golden Hive)

Wooden Version

One of the "bee-natural" hives we work with at the Melissa Garden is the "one-room-hive" (in German: "Einraumbeute"), which is also called the "golden hive". It is designed to provide the best environment for the development of the "Bien" and to minimize necessary manipulation (more frequent opening of hives may result in a weakening of the "Bien").

Four different elements are part of the new design:
- The entire colony lives in one room (without multiple hives and frame levels)
- The hive comes with tall frames. That size comb sustains the "Bien" and allows the development of a large brood nest.

- The side window enables the beekeeper to receive information about the cycle/status of the "Bien" without having to open the hive. A small size "indicator comb" can be build alongside the viewing window.
- The dimensions of the one-room hive are set according to the "golden mean". It is a universal principle within all forming forces in nature and is found in art, architecture and ancient philosophy. It's also called the "divine proportion".

It was designed by Mellifera e.V., the German holistic beekeeping association. The "Golden Hive" provides an environment for bees that is closer to their natural gestalt. It gives the bees the space to build natural comb with greater depth than regular bee hives. The brood nest is a protected space, and honey can be received from the sides. This hive contains 20 frames and is not supered. The comb surface area equals the frames of two regular deep and one medium Langstroth hive bodies. It has the typical screened bottom board for varroa monitoring, and uses follower boards to support changing bee populations throughout the season. A wax cloth lays on top of the frames and provides further options for protection the inner climate of the "Bien".

"Weissenseifener Haengekorb" (Round Skep Hive)

Round Skep Hive

The "Weissenseifener Haengekorb" was designed by the German sculptor Guenther Mancke. The form and shape of the hive are created according to natural/wild bee hives. The "Bien" as "an organic interpretation of an individual" (Tautz) was the blueprint for the design. Already through his outer shape it reveals the nature of the bees colony – as if the egg shaped skep would be the outer shell or skin of this living being. The inner shape allows bees to unfold their own natural gestalt, in harmony with their instinctual life forces. The "Haengekorb" is made out of rye straw and has nine, half moon shaped, arched, movable frames. Comb is built naturally and can be almost

2 feet deep. Supering is possible while fully protecting the integrity of the brood nest. The entrance is located at the bottom of the hive.

Top Bar Hive

Top Bar Hive

It provides all the features of a natural comb bee hive. Top bars fully cover the upper opening of the hive, with initial comb guidance on the lower bar side. Top bar hives are used in many different cultures. We are introducing new versions of the top bar hive this coming year, which will use "bee-natural" hive proportions and will provide more space for larger comb creation. Top bar hives can be built easily with some basic materials.

Langstroth hives with natural comb
All of the Langstroth hives at "The Melissa Garden" allow bees to live on natural comb. Deep hive bodies are added from below (no supering). Bees over-winter on their own food and only true surplus honey is harvested.

Principles of Holistic Beekeeping

Interconnectedness
There is no single bee – as there is no single human being. It's a product of a limited world view. The single bee is only one individual part of the bigger entity of the entire bee hive. The "Bien" is what Tautz calls an "organic interpretation of an individual". This can serve as a beautiful metaphor or mirror for our own existence: That we are just one individual part of a bigger entity: The earth's ecosystem and the entire universe.

Shift Paradigms

Bees are an indicator species, reflecting the health and status of our life environment as well as the interdependency and interconnectedness of all life on earth. Traditional beekeeping and farming understood and acknowledged the natural life forces of the bees. Modern beekeeping and farming practices have lost this ancient knowledge and this loss has taken its toll on the bees on multiple levels. We have to shift paradigms in order to save the bees.

"Bien"

The concept of the "Bien" describes the undividable entity of the hive. The whole is one organism and the hive is more than the sum of the individual parts. Thousands of bees are integrated into a higher-order entity, one whose abilities far transcend those of the individual bee. "The consciousness of the beehive (not of the individual bees) is of a very high nature" (Rudolf Steiner). Their communication and networking capacities, non hierarchical decision processes and an understanding of service to the greater web of life, which the individual being (bee) is part of, are pointing to a higher level of development and awareness. And such, the bees are a vital part of human culture and an inspiration to the soul. Being in touch with the "Bien" also means to reach out to the flowering world. As bee-keepers we are becoming "flower-keepers" and stewards of the earth as well.

Comb

The comb co-evolved with the bees as a part of the bee itself. Wax comb is the biggest inner organ of the "Bien". Bees spend 90% of their life on the comb. They create the wax out of their own body – no other insect is able to do this. The comb is home, womb, pantry, (external) skeleton, sense organ, nerve system, memory organ and immune system. The "comb-wide-web" provides a means of communication on multiple levels: dance, vibration, chemical marking. The dance floor is marked with some bee pheromones and other still unknown substances. The comb is a controlled environment. As the interior milieu, it becomes part of the "Bien". Therefore, it is essential to allow bees to build their own natural comb and to give them the freedom to express their instinctual life forces. Natural comb is essential when we want to support the bees in a time of ecological challenges. It is their birthright.

It is easy to let bees build natural comb. Since bees build according to gravity, hives need to be leveled and frames need to provide some initial guidance, such as beads of wax across the bottom of the top bar, or one-inch strips of beeswax foundation placed for the bees to start building from. Tapered frames may be used as well, since bees draw naturally comb from thin edges. If you have questions with regards to natural comb management, please contact us for assistance.

Nest density – the landscape becomes the apiarium

Following the movement of the "Bien", swarms aim to settle further away from their mother colony. The natural distribution of nest sites vary according to climate and local flora. It is a natural instinct of the "Bien" to leave home and journey into the landscape. It not only reflects a consideration for forage, but it also serves the health of the bees by favoring vertical over horizontal transmission of so called pathogens. The former leads to a lower virulence of diseases where the latter leads to an imbalance of relationships, in particular within the symbiotic life between the "Bien" and its microorganisms and parasites like varroa. Therefore the landscape shall become the apiarium again. Hives are set up individually and spaced at least 400 yards from each other. Various hives are set up on tables and in trees within the boundaries of the Melissa Garden. Empty hives will be "seeded" by swarms only, whether feral or from Melissa Garden hives. We hope to help restore the natural web of life this way on various levels.

Parallels between "Bien" and mammals

There are interesting parallels between bees and mammals. Both have low reproduction rates in common. Mammals raise their offspring with mother's milk and nursing bees use "sister's milk," which is produced in special glands. Wherein mammals provide a uterus, the brood nest of the bees has similar characteristics as a "social" uterus. Body temperatures are 98 degrees F for humans and 95 degrees F for bees, which is very close.

Retention of nest scent and heat

The hive is not an external dwelling of the "Bien". When we open a bee hive, we are entering an ecological system, or even a being's body. Heat, humidity, light, draft, the entire self awareness, the immune system and the sense of integrity of the "Bien" are challenged and affected! Johann Thür called the internal conditions "the element of life, the retention of nest scent and heat", which are part of the immune system and important for its well being (studies have shown that more frequent opening of hives caused higher damage due the hive beetle).

Approach

When opening a hive, how do we approach the "Bien"? It seems we are conditioned to reach for beekeeping tools like a veil, gloves and smoke. Our "armor" may make it challenging to stay open to the mood of the bees and it may be more difficult to extend our empathy. With the protection in place, we more likely penetrate realms of highest vulnerability, without even noticing. We may get out of touch with what is right in front of us. To re-evaluate our precautions and body protections (which are important), we may open up to a new approach of respectful encountering the "Bien". We can learn to listen and how to move and "dance" with the bees. This does not mean not to protect ourselves. The issue is rather not to become "blind" through our "beekeeping tools". Smoke is another stress factor in beekeeping, and with all our senses open, the smoker can stay cold for most of the time. Experiment with it and go with your comfort level. Then a different path of beekeeping opens up.

Excerpts from Demeter Standards for Beekeeping and Hive Products

The Melissa Garden is utilizing biodynamic agriculture methods and will seek Demeter certification. By extension, our beekeeper, Michael Thiele, uses biodynamic beekeeping methods.

- With the exception of fixings, roof coverings and wire meshing, hives must be built entirely of natural materials such as wood, straw or clay. The inside of the hive may only be treated with beeswax and propolis. Only natural, ecologically safe and non-synthetic wood preservatives may be applied to the hive exterior.
- Swarming is the natural way to increase the number of bee colonies and is the only permitted means for increasing colony numbers.
- The system of management cannot rely on the continual introduction of colonies, swarms and queens from elsewhere. Clipping the wings of queens is prohibited. Multiple and routine uniting of colonies as well as systematic queen replacement is not permitted.
- A locally adapted breed of bee suited to the landscape should be chosen.
- The comb is integral to the beehive. Therefore all combs should be constructed as natural combs. Natural combs are those constructed by the bees without the help of waxed midribs. Natural combs can be constructed on fixed or movable frames. Strips of beeswax foundation to guide comb building is permitted.
- The brood area naturally enough forms a self-contained unity. Both comb and brood area must be able to grow as the bee colony develops through building more natural comb. The brood chamber and frame size must be so chosen that the brood area can expand organically with the combs and without being obstructed by wood from the frames. Separation barriers are not allowed as integral elements of the management system.
- Honey and blossom pollen are the natural foods for bees. The aim should be to winter them on honey. All pollen substitutes are forbidden.
- A bee colony should be able to correct any occurring imbalances out of its own resources. Measures taken by the Demeter beekeeper

47

should aim to reinforce and maintain its vitality and capacity for self regeneration. The occasional loss of colonies particularly susceptible to certain pests and diseases should be accepted as a necessary part of natural selection.

Further information on alternative hives at gaiabees.com
Books & web links:
"The Buzz about Bees", by Prof. Dr. Juergen Tautz
"Bees", by Rudolf Steiner
"Wisdom of the bees", by Erik Berrevoets
"Toward saving the Honey Bee", by Guenther Hauk
"Bees and Honey: From Flower to Jar", by Michael Weiler
"The sacred Bee" by Hilda M. Ransome
"The Shamanic way of the bees", by Simon Buxton
"The lost language of plants", by Stephen Harrod Buhner
"Fruitless Fall", by Rowan Jacobsen
www.spikenardfarm.org
www.demeter.net/
www.attra.ncat.org/attra-pub/nativebee.html
www.partnersforsustainablepollination.org/
www.biobees.com
www.beeguardian.org
www.mellifera.de

GREAT, ARCHITECT. IT LOOKS SPACIOUS AND WELCOMING. JUST THE WAY WE DREAMED ABOUT IT.

www.dosisdiarias.com

THE STORY OF THE YEAR IN PICTURES A TWO YEAR BAN ON NEONICOTINOIDS

●

Early in the year it seemed that the banning of neonicotinoids on crops on which bees forage might not go ahead, despite the recommendation of the European Food Safety Authority which strongly believed that the insecticides were responsible for the deaths of thousands of colonies which are so important for pollination. The Environment Minister, Owen Patterson, openly opposed the proposal, though fortunately at the second vote, the ban was put in place for two years, starting from December 2013.

It is believed that millions of bees are killed each year through the use of neonicotinoid insecticides. The insidious nature of the problem is that usually the dead bodies of the bees are not found at the hive entrance, for the sub-lethal effects of the poisons upset the nervous system of the bees, affecting their ability to orientate so they die away from their colony.

THE EU VOTE ON THE BANNING OF THREE NEONICOTINOIDS, 29th APRIL 2013 FOR TWO YEARS

Against the ban were: the UK, Hungary, Czech Republic, Italy, Romania, Portugal, Slovakia and Austria. Finland, Ireland, Lithuania and Greece abstained. Germany, Spain, Belgium, Bulgaria, Luxembourg, Denmark, Estonia, the Netherlands Latvia, Malta, Poland, Cyprus, Slovenia Sweden and France voted in favour.

AGAINST   ABSTAINED   FOR

How the voting went at the second reading of the proposal.

Throughout the campaign many organisations concerned with the environment, wildlife and conservation supported the beekeepers. Unfortunately, there were influential people within beekeeping who agreed with the government's stance.

The campaigns against the use of neonicotinoids and those who manufacture them has made the general public more than ever aware of just how much we rely on both honeybees and other wild pollinators for the majority of our food crops.

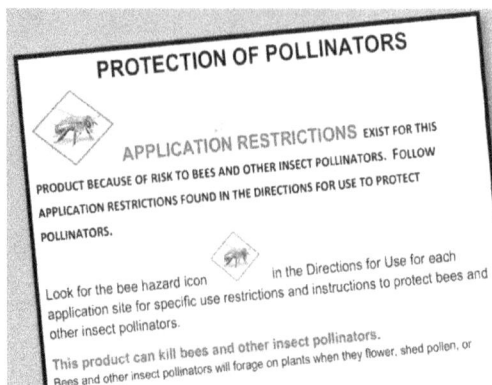

PROTECTION OF POLLINATORS

APPLICATION RESTRICTIONS EXIST FOR THIS PRODUCT BECAUSE OF RISK TO BEES AND OTHER INSECT POLLINATORS. FOLLOW APPLICATION RESTRICTIONS FOUND IN THE DIRECTIONS FOR USE TO PROTECT POLLINATORS.

Look for the bee hazard icon in the Directions for Use for each application site for specific use restrictions and instructions to protect bees and other insect pollinators.

This product can kill bees and other insect pollinators. Bees and other insect pollinators will forage on plants when they flower, shed pollen, or

Beekeepers and organisations in the United States want similar action to protect pollinating insects. The Environmental Protection Agency has responded by merely ensuring that pesticide labels give more information about the damage that they can do.

Do you think a label can help the Bees?

The EPA does.

Bee Against Monsanto

This response from the EPA has been met with disdain by all involved in the campaigns against the pesticide producers.

Meanwhile, bee colonies are likely to continue to suffer.

BREAKING NEWS:
SYNGENTA & BAYER SUE THE
EUROPEAN UNION OVER PESTICIDES BAN,

THINK THEIR PROFITS ARE
MORE IMPORTANT THAN THE BEES.

www.gmofreeusa.org

The chemical companies are responding by suing the European Union.

54

WORD SEARCH

THE GRID BELOW CONTAINS THE NAMES OF 21 FAMOUS PEOPLE WHO ARE/HAVE ALSO BEEN BEEKEEPERS. THEIR NAMES MAY BE FOUND BY SEARCHING THE GRID HORIZONTALLY, VERTICALLY AND DIAGONALLY, BOTH BACKWARDS AND FORWARDS.

```
Y  L  L  E  W  O  P  N  E  D  A  B  D  R  O  L  T  W
G  X  U  O  R  E  H  T  L  U  A  P  S  Q  Y  R  N  Y
A  B  M  L  B  J  N  L  L  Y  S  T  U  Z  D  Q  V  H
D  I  C  P  B  O  L  R  I  U  I  E  T  A  P  A  O  P
N  O  T  E  V  H  R  Y  V  L  H  F  I  W  T  I  H  I
O  T  Y  T  G  N  C  R  L  T  R  C  R  V  G  N  T  Y
F  Z  P  E  R  W  X  Q  X  H  Y  C  C  I  P  O  A  S
Y  V  V  R  E  H  Z  F  W  O  G  P  O  K  P  S  L  I
R  N  U  F  G  I  R  A  P  M  L  P  M  T  Y  R  P  F
N  S  M  O  O  T  F  H  P  A  C  A  E  O  T  E  A  G
E  A  K  N  R  T  L  N  K  S  R  R  D  R  H  F  I  V
H  Z  O  D  M  I  E  Y  I  E  D  T  O  Y  A  F  V  V
K  I  A  A  E  E  O  U  N  D  C  N  W  U  G  E  L  X
S  A  C  V  N  R  T  Q  P  I  T  O  H  S  O  J  Y  Q
L  S  W  J  D  R  O  Y  V  S  I  V  G  C  R  S  S  V
P  K  T  M  E  P  L  U  F  O  U  A  I  H  A  A  T  U
L  E  F  T  L  C  S  Q  F  N  Q  I  K  E  S  M  F  X
O  J  Y  K  Q  L  T  E  X  B  O  R  O  N  G  O  L  E
V  I  R  G  I  L  O  L  F  B  P  A  H  K  W  H  D  G
S  A  Y  P  L  A  Y  V  C  L  U  M  D  O  E  T  G  M
```

ANSWERS ON PAGE 202

Selections from
The Historical Honeybee Articles

*I here present thee with a hive of bees, laden some with wax,
and some with honey. Fear not to approach! There are no
Wasps, there are no Hornets here. If some wanton Bee should
chance to buzz about thine ears, stand thy ground and hold
thy hands: there's none will sting thee if thy strike not first.
If any do, she hath honey in her bag will cure thee too.*

Francis Quarles

This quote by the 17th Century poet appears on the welcome page of
Facebook's The Historical Honeybee Articles. The site owner is keen to
share the images with other beekeepers and welcomes the addition of more
illustrations to its pages.

DIARY & CALENDAR

14

- PART II -

*SR (SUNRISE) SS (SUNSET) FOR LONDON UK.

JANUARY

The English poet Francis Quarles, (1592 - 1644) said:
Flatter not thyself in thy faith in God if thou hast not charity for thy neighbor."

"Charity is a naked child, giving
honey to a bee without wings;
naked, because excuseless and
simple; a child, because tender
and growing: giving honey, because
honey is pleasant and comfortable: ...
to a bee, because a bee is laborious
and deserving: without wings,
because helpless, and wanting.
If thou denies to such thou kills a
bee; if thou gives to other than such,
thou preserves a drone."

- Francis Quarles, Enchiridion 1640

DAY	JANUARY 2014 FORAGE	TEMP		WIND		CL'D	RAIN	1	2	3
		MIN	MAX	DIR	B.S			HIVE WEIGHT		
1										
2										
3										
4										
5										
6										
7										
8										
9										
10										
11										
12										
13										
14										
15										
16										
17										
18										
19										
20										
21										
22										
23										
24										
25										
26										
27										
28										
29										
30										
31										

JAN 14

1,WE ○
NEW YEAR'S DAY

2,TH

3,FR

4,SA
SR:08:06, SS:16:05

5,SU

6,MO
EPIPHANY

7,TU
ORTHODOX CHRISTMAS DAY

8,WE

9,TH

10,FR

11,SA
SR:08:02, SS:16:15

12,SU

13,MO

14,TU

15,WE

16,TH ●	**24,FR**
17,FR	25,SA SR:07:49, SS:16:37 BURNS' NIGHT
18,SA SR:07:57, SS:16:25	26,SU
19,SU	**27,MO**
20,MO	**28,TU**
21,TU	**29,WE**
22,WE	**30,TH** ○
23,TH	**31,FR** CHINESE NEW YEAR

FEBRUARY

Jupiter and the Bee

In days of yore, when the world was young, a Bee that had stored her combs with a bountiful harvest, flew up to heaven to present as a sacrifice an offering of honey. Jupiter was so delighted with the gift, that he promised to give her whatsoever she should ask for. She therefore besought him, saying, "O glorious Jove, maker and master of me, poor Bee, give thy servant a sting, that when any one approaches my hive to take the honey, I may kill him on the spot." Jupiter, out of love to man, was angry at her request, and thus answered her: "Your prayer shall not be granted in the way you wish, but the sting which you ask for you shall have; and when any one comes to take away your honey and you attack him, the wound shall be fatal not to him but to you, for your life shall go with your sting."

Moral:
He that prays harm for his neighbor, begs a curse upon himself.

DAY	FEBRUARY 2014 FORAGE	TEMP MIN	MAX	WIND DIR	B.S	CL'D	RAIN	1	2	3
								\multicolumn{3}{c}{HIVE WEIGHT}		
1										
2										
3										
4										
5										
6										
7										
8										
9										
10										
11										
12										
13										
14										
15										
16										
17										
18										
19										
20										
21										
22										
23										
24										
25										
26										
27										
28										

FEB14

	8,SA SR:07:27, SS:17:03
1,SA SR:07:39, SS:16:50	**9,SU**
2,SU CANDLEMAS	**10,MO**
3,MO	**11,TU**
4,TU	**12,WE**
5,WE	**13,TH**
6,TH	**14,FR** ● VALENTINE'S DAY
7,FR	**15,SA** SR:07:15, SS:17:15

16,SU	24,MO
17,MO	25,TU
18,TU	26,WE
19,WE	27,TH
20,TH	28,FR
21,FR	
22,SA SR:07:01, SS:17:28	
23,SU	

Mother Bee NURSERY RHYMES
By M.G.P. (Mother Goose Plagiarized)

"Where are you going, my pretty maid?"
"I'm going to the beeyard, Sir," she said,
"May I go with you, my pretty maid?"
"If you'll lift the hives, kind sir," she said.

"What is your fortune, my pretty maid?"
"My bees are my fortune, Sir," she said.
"Then I shall marry you, my pretty maid!"
"You'll be stung if you do, kind Sir," she said."

DAY	MARCH 2014 FORAGE	TEMP		WIND		CL'D	RAIN	1	2	3
		MIN	MAX	DIR	B.S			\multicolumn{3}{c}{HIVE WEIGHT}		
1										
2										
3										
4										
5										
6										
7										
8										
9										
10										
11										
12										
13										
14										
15										
16										
17										
18										
19										
20										
21										
22										
23										
24										
25										
26										
27										
28										
29										
30										
31										

MAR14

	8,SA SR:06:31, SS:17:53 UBKA ANNUAL CONFERENCE, GREENMOUNT CAMPUS, ANTRIM
1,SA ○ SR:06:46, SS:17:41 ST DAVID'S DAY BEEKEEPING TRADE SHOW, STONELEIGH PARK	**9,SU**
2,SU	**10,MO**
3,MO	**11,TU**
4,TU SHROVE TUESDAY	**12,WE**
5,WE ASH WEDNESDAY	**13,TH**
6,TH	**14,FR**
7,FR UBKA ANNUAL CONFERENCE, GREENMOUNT CAMPUS, ANTRIM	**15,SA** SR:06:15, SS:18:05

16,SU ●	24,MO
17,MO ST PATRICK'S DAY	25,TU
18,TU	26,WE
19,WE	27,TH
20,TH SPRING EQUINOX	28,FR
21,FR	29,SA
22,SA SR:05:59, SS:18:17	30,SU DAYLIGHT SAVING TIME BEGINS MOTHERING SUNDAY
23,SU	31,MO ○

EVERY GOOD MOTHER SHOULD BE THE HONORED QUEEN OF A HAPPY FAMILY.

Happy Mothers Day!

For Mother's Day

Langstroth placed this image and text on the inside cover of 'The Hive and the Honeybee' as a reminder to us all that every happy family grows from the love and affection which a mother provides. And every good mother should be the honored queen of a happy family.

"… her movements are measured and majestic, as she moves in the hive the other bees form a circle round her, none venturing to turn their backs upon her, but all anxious to show that respect and attention due to her rank and station. Whenever in the exercise of her sovereign will the queen wishes to travel amongst her subjects, she experiences no inconvenience from overcrowding; although the part of the hive to which she is journeying may be the most populous, way is immediately made, the common bees tumbling over each other to get out of her way, so great is their anxiety not to interfere with the royal progress." Alfred Neighbour, 'The apiary; or, Bees, bee-hives and bee culture' 1865.

DAY	APRIL 2014 FORAGE	TEMP MIN	TEMP MAX	WIND DIR	WIND B.S	CL'D	RAIN	1	2	3
								HIVE WEIGHT		
1										
2										
3										
4										
5										
6										
7										
8										
9										
10										
11										
12										
13										
14										
15										
16										
17										
18										
19										
20										
21										
22										
23										
24										
25										
26										
27										
28										
29										
30										

APR14

8,TU

1,TU

9,WE

2,WE

10,TH

3,TH

11,FR

4,FR
BBKA SPRING CONVENTION
HARPER ADAMS COLLEGE, HEREFORD

12,SA
SR:06:12, SS:19:52

5,SA
SR:06:27, SS:19:40
BBKA SPRING CONVENTION
HARPER ADAMS COLLEGE, HEREFORD

13,SU
PALM SUNDAY

6,SU
BBKA SPRING CONVENTION
HARPER ADAMS COLLEGE, HEREFORD

14,MO

7,MO

15,TU ●
TOTAL LUNAR ECLIPSE

16,WE	**24,TH**
17,TH MAUNDY THURSDAY	**25,FR**
18,FR GOOD FRIDAY/ORTHODOX GOOD FRIDAY	26,SA SR:05:42, SS:20:15
19,SA SR:05:57, SS:20:04 HOLY SATURDAY/ ORTHODOX HOLY SATURDAY	27,SU
20,SU EASTER SUNDAY/ORTHODOX EASTER	**28,MO**
21,MO EASTER MONDAY	**29, TU** ○
22,TU	**30, WE**
23,WE ST GEORGE'S DAY	

MAY

The Hornet and the Bees

Jean de La Fontaine, (1621-1695), French author who lived at
Château Thierry and Paris. Though his stories are well known only in France,
his fables are read throughout the world. They are fresh, vivid, and artistic, pleasing
young and old alike to a remarkable degree.

Illustration from The Hornet and the Bees, by J J Grandville.
From Fables de La Fontaine, by Jean de La Fontaine, Paris, 1855.

DAY	MAY 2014 FORAGE	TEMP		WIND		CL'D	RAIN	1	2	3
		MIN	MAX	DIR	B.S			HIVE WEIGHT		
1										
2										
3										
4										
5										
6										
7										
8										
9										
10										
11										
12										
13										
14										
15										
16										
17										
18										
19										
20										
21										
22										
23										
24										
25										
26										
27										
28										
29										
30										
31										

MAY 14

	8,TH
1,TH	**9,FR**
2,FR	10,SA SR:05:17, SS:20:38
3,SA SR:05:29, SS:20:27	11,SU
4,SU	**12,MO**
5,MO EARLY MAY BANK HOLIDAY	**13,TU**
6,TU	**14,WE** ●
7,WE	**15,TH**

16,FR	**24,SA** SR:04:57, SS:20:58
17,SA SR:05:06, SS:20:49	**25,SU**
18,SU	**26,MO** SPRING BANK HOLIDAY
19,MO	**27,TU**
20,TU	**28,WE** ○
21,WE	**29,TH** ASCENSION DAY
22,TH	**30,FR**
23,FR	**31,SA** SR:04:50, SS:21:07

JUNE

For Father's Day
Fathers of the Bee People (Edward Bevan -1843)

The drones or males are at once her majesty's nobles and husbands, dividing with her the administrative care of the State, the official trusts, and the parental functions. They are the office-holders and politicians; having, in general, little to do but to buz about royalty, pay their court, eat the fat and the sweat of the land, and talk politics. Their number varies with the strength of the hive, from fifteen hundred to two thousand. They seem to be, for nobles and husbands, rather unwarlike; for they possess no stings. On the whole, as they neither fight nor work, but only make love, they must have rather an easy time of it. Still, as we do not choose to injure anybody's character, we feel bound to say that, if they mix not in the ordinary tasks of the operative Bees, it is the fault of nature, and not theirs: for she has furnished them with neither the sort of trowel to the jaws, with which the workers manage the wax, nor the baskets to the legs, in which they collect their fragrant spoil from the flowers. They labor not, then, because they have higher functions to perform, of a far loftier consequence to the public weal. And their wise and just fellow-citizens, content that each order in the State should discharge its appropriate duty, murmur not, nor stigmatize them as non-producers, nor rail nor roar at them as aristocrats; but recognize their utility in the peculiar part which has been assigned them of the public business, and submit with cheerfulness to their exemption from inferior tasks, inappropriate as well as impossible to these general fathers of the Bee people.

'Bevan on the Bee', Second Notice, 1843,
The Gentleman's Magazine, and Historical Chronicle,
Volume 97, Part 1 1827, page 608

DAY	JUNE 2014 FORAGE	TEMP		WIND		CL'D	RAIN	1	2	3
		MIN	MAX	DIR	B.S			HIVE WEIGHT		
1										
2										
3										
4										
5										
6										
7										
8										
9										
10										
11										
12										
13										
14										
15										
16										
17										
18										
19										
20										
21										
22										
23										
24										
25										
26										
27										
28										
29										
30										

JUN14

1,SU	**9,MO** WHIT MONDAY
2,MO	**10,TU**
3,TU	**11,WE**
4,WE	**12,TH**
5,TH	**13,FR** ●
6,FR	14,SA SR:04:43, SS:21:19
7,SA SR:04:45, SS:21:14	15,SU FATHER'S DAY

16,MO	**24,TU**
17,TU	**25,WE**
18,WE	**26,TH**
19,TH ◉	**27,FR** ○
20,FR	28,SA SR:04:46, SS:21:22
21,SA SR:04:43, SS:21:21 SUMMER SOLSTICE	29,SU RAMADAN BEGINS
22,SU	**30,MO**
23,MO	

The Legend of St. Peter and the Bees

The legend relates how St Peter complained to our Lord that the innocent were
punished with the guilty. Our Lord made no answer, but shortly after commanded
St. Peter to pick up a piece of honey-comb filled with bees, and put it in the bosom of
his dress. One of the bees stung him, and St. Peter in his anger killed them all,
and when the Lord rebuked him, excused himself by saying:
"How could I tell ...among so many bees which one stung me?"
The Lord answered: "Am I wrong then, when I punish men likewise?"
"Chianci lu giustu pri lu piccaturi" -"The Just suffers for the Sinner" .

DAY	JULY 2014 FORAGE	TEMP		WIND		CL'D	RAIN	1	2	3
		MIN	MAX	DIR	B.S			HIVE WEIGHT		
1										
2										
3										
4										
5										
6										
7										
8										
9										
10										
11										
12										
13										
14										
15										
16										
17										
18										
19										
20										
21										
22										
23										
24										
25										
26										
27										
28										
29										
30										
31										

JUL14

	8,TU
1,TU	**9,WE**
2,WE	**10,TH**
3,TH	**11,FR**
4,FR	12,SA ● SR:04:57, SS:21:14
5,SA SR:04:51, SS:21:19	13,SU
6,SU	**14,MO**
7,MO	**15,TU**

16,WE	**24,TH**
17,TH	**25,FR**
18,FR	26,SA ○ SR: 05:15, SS:20:58
19,SA SR:05:06, SS:21:07	27,SU GORMANSTON CONFERENCE, EIRE
20,SU	**28,MO** GORMANSTON CONFERENCE, EIRE
21,MO	**29, TU** GORMANSTON CONFERENCE, EIRE
22,TU	**30, WE** GORMANSTON CONFERENCE, EIRE
23,WE	**31, TH** GORMANSTON CONFERENCE, EIRE

AUGUST

https://www.facebook.com/Historical.Honeybee.Articles

The Beekeepers and the Birdnester

Circa 1568
by Pieter Brueghel the Elder

Pieter Brueghel the Elder was a Flemish Renaissance painter and printmaker known for his landscapes and peasant scenes. The drawing depicts beekeepers performing some type of early season beekeeping, - perhaps setting up an apiary with newly-hived swarms. The person climbing the tree is a 'birdnester'. A birdnester is a person who collects hunts and gathers the nests of birds in order to get the birds eggs, and the act of hunting for these nests is referred to as birdnesting. Birdnesting is typically most productive during the time of the season in which bees are swarming, so we assume here that the drawing by Brueghel may be a depiction of 16th century beekeepers eagerly setting up an apiary with newly hived swarms.

DAY	AUGUST 2014 FORAGE	TEMP MIN	MAX	WIND DIR	B.S	CL'D	RAIN	1	2	3
								HIVE WEIGHT		
1										
2										
3										
4										
5										
6										
7										
8										
9										
10										
11										
12										
13										
14										
15										
16										
17										
18										
19										
20										
21										
22										
23										
24										
25										
26										
27										
28										
29										
30										
31										

AUG14

	8,FR
1,FR GORMANSTON CONFERENCE, EIRE	**9,SA** SR:05:36, SS:20:35
2,SA SR:05:25, SS:20:47	**10,SU** ●
3,SU	**11,MO**
4,MO BANK HOLIDAY (SCOTLAND)	**12,TU** PERSEIDS METEOR SHOWER GROUSE SHOOTING BEGINS - MOVE BEES TO THE HEATHER
5,TU	**13,WE** PERSEIDS METEOR SHOWER
6,WE	**14,TH**
7,TH	**15,FR** ASSUMPTION OF MARY (ORTHODOX)

16,SA SR:05:47, SS:20:21	24,SU ST BARTHOLOMEW'S DAY - TRADITIONAL TIME FOR EXTRACTING HONEY
17,SU	**25,MO** ○ SUMMER BANK HOLIDAY
18,MO	**26,TU**
19,TU	**27,WE**
20,WE	**28,TH**
21,TH	**29,FR**
22,FR	30,SA SR:06:10, SS:19:52
23,SA SR:05:58, SS:20:07	31,SU

SEPTEMBER

"Praise of the Bees"
Troia Roll, Exultet. Troia Roll, Exultet. Cathedral of Troia (Troja, Apulia, Italy).
Circa 1175

Illuminated illustration depicting bees carrying nectar from flowers to
wooden plank hives.

DAY	SEPTEMBER 2014 FORAGE	TEMP		WIND		CL'D	RAIN	1	2	3
		MIN	MAX	DIR	B.S			HIVE WEIGHT		
1										
2										
3										
4										
5										
6										
7										
8										
9										
10										
11										
12										
13										
14										
15										
16										
17										
18										
19										
20										
21										
22										
23										
24										
25										
26										
27										
28										
29										
30										

SEP14

	8,MO
1,MO	9,TU ●
2,TU	10,WE
3,WE	11,TH
4,TH	12,FR
5,FR	13,SA SR:06:32, SS:19:20
6,SA SR:06:21, SS:19:36	14,SU
7,SU	15,MO

16,TU	**24,WE** ○
17,WE	**25,TH**
18,TH	**26,FR**
19,FR	27,SA SR:06:54, SS:18:48
20,SA SR:06:43, SS:19:04	28,SU
21,SU	**29,MO**
22,MO	**30,TU**
23,TU AUTUMN EQUINOX	

OCTOBER

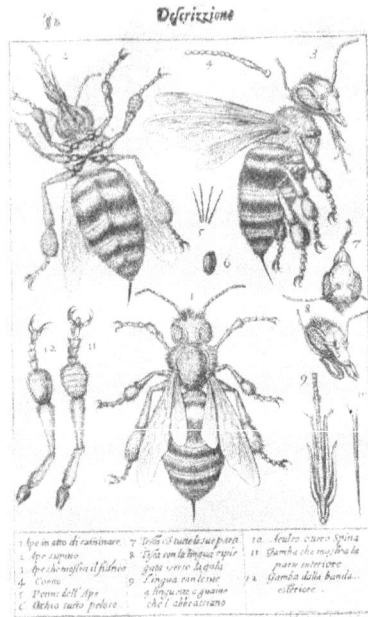

Engraved Plate Depicting the Honeybee as
Seen Through a New Device Called the Microscope.

The Apiarium (Rome, 1625) was a gift of the Lynx to the new pope. Galileo adapted the telescope into a new instrument, named a microscope by a member of the Lynx. Just as Galileo's telescope brought near the Moon and stars, so could the eyes of the Lynx see the secrets of the small, portraying structures of the bee never seen before. In the Apiarium, the first publication of observations made with a microscope, Federico Cesi (1585-1630) and Francesco Stelluti (1577-1651) studied the anatomy of the honeybee.

In a work of the same time, Stelluti published drawings. On the title page, a remarkable plate displaying the fine anatomical structures of honeybees, arranged in the pattern of the Barberini.

DAY	OCTOBER 2014 FORAGE	TEMP		WIND		CL'D	RAIN	1	2	3
		MIN	MAX	DIR	B.S			HIVE WEIGHT		
1										
2										
3										
4										
5										
6										
7										
8										
9										
10										
11										
12										
13										
14										
15										
16										
17										
18										
19										
20										
21										
22										
23										
24										
25										
26										
27										
28										
29										
30										
31										

OCT14

	8,WE ● TOTAL LUNAR ECLIPSE
1,WE	**9,TH**
2,TH ◐	**10,FR** SR:07:16, SS:18:18
3,FR	11,SA
4,SA SR:07:06, SS:18:32 YOM KIPPUR FEAST OF ST FRANCIS OF ASSISI	12,SU
5,SU	**13,MO**
6,MO	**14,TU**
7,TU	**15,WE**

16,TH	**24,FR** NATIONAL HONEY SHOW
17,FR	25,SA SR:07:41, SS:17:47 NATIONAL HONEY SHOW
18,SA SR:07:29, SS:18:01	26,SU DAYLIGHT SAVING TIME ENDS
19,SU	**27,MO**
20,MO	**28,TU**
21,TU	**29,WE**
22,WE	**30,TH**
23,TH ○ NATIONAL HONEY SHOW	**31,FR** ●

NOVEMBER

An illustration of bee-keeping from the 1697 edition of
The Georgics of Virgil, part IV, translated by John Dryden.

Virgil was a classical Roman poet, best known for three major works —the Eclogues,
the Georgics, and the Aeneid —although several minor poems are also attributed to
him. The son of a farmer, Virgil came to be regarded as one of Rome's greatest poets.

The poet Virgil is beloved among apiarists for his didactic poem titled The Georgics,
which gave instruction in the methods of running a farm. The Georgics focus
respectively in raising crops and trees in books I and II, livestock and horses in book
III, and beekeeping and the qualities of bees in book IV.

DAY	NOVEMBER 2014 FORAGE	TEMP MIN	TEMP MAX	WIND DIR	WIND B.S	CL'D	RAIN	1	2	3
								HIVE WEIGHT		
1										
2										
3										
4										
5										
6										
7										
8										
9										
10										
11										
12										
13										
14										
15										
16										
17										
18										
19										
20										
21										
22										
23										
24										
25										
26										
27										
28										
29										
30										

NOV14

	8,SU SR:07:06, SS:16:22
1,SU SR:06:54, SS:16:34 ALL SAINTS' DAY	**9,MO**
2,MO ALL SOULS' DAY	**10,TU**
3,TU	**11,WE**
4,WE	**12,TH**
5,TH GUY FAWKES DAY	**13,FR**
6,FR ◉	**14,SA**
7,SA	**15,SU** SR: 07:18, SS:16:11

16,MO	**24,TU**
17,TU	**25,WE**
18,WE	**26,TH**
19,TH	**27,FR**
20,FR	28,SA
21,SA	29,SU SR:07:41, SS:15:56
22,SU ○ SR:07:30, SS:16:03	**30,MO** ST ANDREW'S DAY
23,MO	

DECEMBER

St Ambrose of Milan
Patron Saint of Beekeepers

The legend is that as an infant, a swarm of bees settled on his face while he lay in his cradle, leaving behind a drop of honey. His father considered this a sign of his future eloquence and honeyed tongue. For this reason, bees and beehives often appear as symbols of Ambrose and he is known throughout the world as being the Patron Saint of Beekeepers, his feast day being on the 7th December.

DAY	DECEMBER 2014 FORAGE	TEMP MIN	TEMP MAX	WIND DIR	WIND B.S	CL'D	RAIN	1	2	3
								\multicolumn{3}{c}{HIVE WEIGHT}		
1										
2										
3										
4										
5										
6										
7										
8										
9										
10										
11										
12										
13										
14										
15										
16										
17										
18										
19										
20										
21										
22										
23										
24										
25										
26										
27										
28										
29										
30										
31										

DEC14

	8,MO
1,MO	**9,TU** RAMADAN BEGINS
2,TU	**10,WE**
3,WE	**11,TH**
4,TH	**12,FR**
5,FR	13,SA SR:07:58, SS:15:51
6,SA ● SR:07:50, SS:15:53	14,SU
7,SU	**15,MO**

16,TU	**24,WE** LAST DAY OF HANNUKAH
17,WE FIRST DAY OF HANNUKAH	**25,TH** CHRISTMAS DAY
18,TH	**26,FR** BOXING DAY
19,FR	27,SA SR:08:06, SS:15:57
20,SA SR:08:03, SS:15:53	28,SU
21,SU WINTER SOLSTICE	**29,MO**
22,MO ○	**30,TU**
23,TU	**31,WE**

Hive/ Q NO.	Year Q Raised	Frames of Brood Autumn 2013	Combs Covered	Honey Stored- Sugar fed Kg	Combs Covered Spring 2014	Frames of Brood Spring 2014	Spring Feeding Kg	Queens Reared	Nuclei
1									
2									
3									
4									
5									
6									
7									
8									
9									
10									
11									
12									
13									
14									
15									
16									
17									
18									
19									
20									
21									
22									
23									
24									

HONEYBEE COLONIES

1								
2								
3								
4								
5								
6								
7								
8								
9								
10								
11								
12								
13								
14								
15								
16								
17								
18								
19								
20								
21								
22								
23								
24								

BEEEKEEPING RECORDS

Number	items	Est. Value	
		£	P
	Stocks of Bees		
	Empty Hives		
	Combs - Deep - Shallow		
	Frames		
	Foundations		
	Honey Extractor		
	Honey Tanks		
	Other items		
	Honey Jars		
	Honey		

JANUARY 2015

S	M	T	W	T	F	S
				1	2	3
4	5	6	7	8	9	10
11	12	13	14	15	16	17
18	19	20	21	22	23	24
25	26	27	28	29	30	31

FEBRUARY 2015

S	M	T	W	T	F	S
1	2	3	4	5	6	7
8	9	10	11	12	13	14
15	16	17	18	19	20	21
22	23	24	25	26	27	28

MARCH 2015

S	M	T	W	T	F	S
1	2	3	4	5	6	7
8	9	10	11	12	13	14
15	16	17	18	19	20	21
22	23	24	25	26	27	28
29	30	31				

APRIL 2015

S	M	T	W	T	F	S
			1	2	3	4
5	6	7	8	9	10	11
12	13	14	15	16	17	18
19	20	21	22	23	24	25
26	27	28	29	30		

MAY 2015

S	M	T	W	T	F	S
					1	2
3	4	5	6	7	8	9
10	11	12	13	14	15	16
17	18	19	20	21	22	23
24	25	26	27	28	29	30
31						

JUNE 2015

S	M	T	W	T	F	S
	1	2	3	4	5	6
7	8	9	10	11	12	13
14	15	16	17	18	19	20
21	22	23	24	25	26	27
28	29	30				

JULY 2015

S	M	T	W	T	F	S
			1	2	3	4
6	7	8	9	10	11	
13	14	15	16	17	18	
20	21	22	23	24	25	
27	28	29	30	31		

AUGUST 2015

S	M	T	W	T	F	S
						1
2	3	4	5	6	7	8
9	10	11	12	13	14	15
16	17	18	19	20	21	22
23	24	25	26	27	28	29
30	31					

SEPTEMBER 2015

S	M	T	W	T	F	S
		1	2	3	4	5
6	7	8	9	10	11	12
13	14	15	16	17	18	19
20	21	22	23	24	25	26
27	28	29	30			

OCTOBER 2015

S	M	T	W	T	F	S
				1	2	
4	5	6	7	8	9	
11	12	13	14	15	16	
18	19	20	21	22	23	
25	26	27	28	29	30	

NOVEMBER 2015

S	M	T	W	T	F	S
1	2	3	4	5	6	7
8	9	10	11	12	13	14
15	16	17	18	19	20	21
22	23	24	25	26	27	28
29	30					

DECEMBER 2015

S	M	T	W	T	F	S
		1	2	3	4	5
6	7	8	9	10	11	12
13	14	15	16	17	18	19
20	21	22	23	24	25	26
27	28	29	30	31		

Ninemaidens
Mead

Award winning mead & honey

visit www.ninemaidensmead.com
or tel. 01209 820939 / 860630

Every effort is made to keep entries up to date but the publishers cannot be held responsible for errors or omissions.

Associations and all other groups listed have been requested (August 2013) to supply updated entries.

Readers who are aware of inaccuracies are asked to send updates to jerry@recordermail.co.uk

DIRECTORY, Associations and Services

DIRECTORY, ASSOCIATIONS AND SERVICES

BEE MAILING

BEEKEEPING MAILING LISTS

http://www.zbee.dircon.co.uk

Beekeeping mailing list services provided by zbee.com http://www.zbee. dircon.co.uk

KENT BEEKEEPERS ASSOCIATION, THE

Name of mailing list: Kentbee-L Serving a possible membership of 400. **Support website:** http://www.kentbee.com Approximately 80 have subscribed. Providing a forum for local branch announcements and news and chat about beekeeping. **To subscribe to Kentbee-L send a message to:** mailserver@zbee.com **Subject field:** You leave this blank it doesn't matter. **In the message body write:** Subscribe Kentbee-L then send the message and await further instructions to complete the subscription process.

NATIONAL HONEY SHOW, THE

Name of mailing list: NHS The National Honey Show is held in October each year in London, the support website http://www.honeyshow.co.uk has more information and schedules, **To subscribe to NHS send a message to:** mailserver@zbee.com, **Subject field:** You leave this blank it doesn't matter., **In the message body write:** Subscribe NHS then send the message and await further instructions to complete the subscription process

BEE IMPROVEMENT & BEE BREEDERS ASSOCIATION, THE (BIBBA)

Name of mailing list: BIBBA-L, Support website http://www.bibba.com/, **To subscribe to BIBBA-L send a message to:** mailserver@zbee.com, **Subject field:** You leave this blank it doesn't matter. **In the message body write:** Subscribe BIBBA-L then send the message and await further instructions to complete the subscription process.

BEE MAILING

✉ ☎

APINET (BEEKEEPING EDUCATION EXTENSION NETWORK)
Name of mailing list: APINETL, Support website n/a, **To subscribe send a message to:** mailserver@zbee.com, **Subject field:** You leave this blank it doesn't matter.
In the message body write: Subscribe APINETL then send the message and await further instructions to complete the subscription process.

BROMLEY & SIDCUP & ORPINGTON BEEKEEPERS ASSOCIATION
Name of mailing list: BBK, **Support website:** http://www.kentbee.com/, **To subscribe to BBK send a message to:** mailserver@zbee.com, **Subject field:** You leave this blank it doesn't matter. **In the message body write:** Subscribe BBK then send the message and await further instructions to complete the subscription process.

THE BRITISH BEEKEEPERS ASOCIATION (BBKA)
Name of mailing list: BBKA, **Support website:** http://www.bbka.org.uk, Private list members only, see members area for joining details.

BEE DISEASES INSURANCE LTD

SECRETARY
Donald Robertson-Adams
Ffosyffin, Ffostrasol,
Llandysul, Ceredigion,
SA44 5JY
07532 336076
secretary@beediseases-
insurance.cso.uk

**TREASURER AND
SCHEME B MANAGER**
Mrs Sharon Blake
Stratton Court,
South Petherton,
Somerset TA13 5LQ
01460 242124
treasurer@beediseases-
insurance.cso.uk

CLAIMS MANAGER
Bernard Diaper
57 Marfield Close,
Walmley,
Sutton Coldfield B76 1YD
07711456932
claims@beediseasesin-
surance.cso.uk

PRESIDENT
Martin Smith
137 Blaguegate Lane
Lathom, Sklemersdale
Lancs WN8 8TX
07831 695732
president@beediseases-
insurance.cso.uk

Bee Diseases Insurance (BDI) provides insurance cover for individual beekeepers, association apiaries and commercial beekeepers alike, against the possibility of their bee equipment being destroyed as a result of a Destruction Order following a visit from an authorised Bee Inspector. .

BDI provides compensation for specified property that may need to be destroyed as a result of American Foul Brood and European Foul Brood.

Also BDI has established a contingency fund capped at £50,000 a year if Small Hive Beetle or Tropilaelaps infestation is found.

Scheme A provides cover for the beekeeper with a total of 39 colonies or less. Cover is obtained by being a member of a Beekeeping Association that is a member of BDI Ltd.

Scheme B provides cover for beekeepers with 40 or more colonies in total. Insurance under this Scheme is on a personal basis and further details can be obtained from the Scheme B Manager.

From 2014 BDI has set up eReturn, allowing on line submission of membership details by treasurers and the generation of electronic receipts for members.

Full details of eReturn and other BDI activities can be found at their web site

www.beediseasesinsurance.co.uk

REMEMBER: DISEASE CAN STRIKE ANY COLONY AT ANY TIME AND IT IS SPREAD THROUGHOUT THE COUNTRY. PROTECT YOUR APIARY, AND OTHER BEEKEEPERS, THROUGH B.D.I.

BEE FARMERS' ASSOCIATION OF THE UNITED KINGDOM

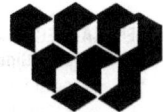

BFA

BEE FARMERS' ASSOCIATION

The BFA represents the professional beekeepers of the UK.

The association is the largest contract pollinator in the UK and our members are responsible for virtually all the migratory pollination. They are expected to have a good degree of competence; membership requires over 40 hives, and sponsorship by a BFA member who knows the applicant as a beekeeper. We have recently introduced a code of conduct which members are expected to observe. In addition we have a significant number of members who get some income from being bee inspectors, responsible for identifying and dealing with notifiable disease.

Business is conducted at twice-yearly regional meetings which pass items up to the main meeting for discussion and voting, and which put forward candidates for the committee.

The BFA is affiliated to the National Farmers Union and The Honey (Packers) Association with whom we work effectively in promoting ecological sensitive farming and in promoting consumer awareness through events such 'National Honey Week' and bulk sales to retail chains.

MEMBERSHIP

Our members are expected to have a good degree of competence.

FULL MEMBERSHIP requires over 40 hives, and sponsorship by a BFA member who knows the applicant as a beekeeper.

ASSOCIATE MEMBERSHIP is a stepping stone to full membership of the BFA for beekeepers with a minimum of 20 hives and who would like to take up commercial or semi-commercial beekeeping.

Membership forms are available from the Membership Secretary, or as a download from our website.

FUNCTIONS

- To monitor and to keep members informed about developments in commercial beekeeping, bee science and UK and EEC legislation.
- Liaison with Farmers, Growers, Contractors, Consumers

CHAIRMAN
Michael Gleeson (FIBKA)
Ballinakill
Enfield
Co.Meath, Ireland.
+353 (0)4695 41433
mgglee@eircom.net

VICE-CHAIR
Margaret Ginman
Hendal House
Hendal Hill
Groombridge
Tunbridge Wells
KENT
TN3 9NT
01892 864 499 /
07795 153 765
margaret.hendal@
btconnect.com

TREASURER,
Doug Isles
Hudnalls Apiary
Balligan Cottage
The Hudnalls
St. Briavels
Lydney
GLOUCESTERSHIRE
GL15 6RT
01594 530807
info@hudnallsapiaries.co.uk

GENERAL SECRETARY,
Margaret Ginman
Hendal House
Hendal Hill
Groombridge
Tunbridge Wells
KENT
TN3 9NT
01892 864 499 /
07795 153 765
margaret.hendal@
btconnect.co

POLLINATION SECRETARY
Alan Hart
61 Fakenham Road,
Great Witchingham,
Norwich,
NORFOLK
NR9 5AE
01603 308911
earlswoodbees@hotmail.
co.uk

RESEARCH AND
ADMINISTRATION
(INCLUDING MEMBERSHIP,
INSURANCE AND BULLETIN)
David Bancalari
Park Farm Barn
Shortthorn Road
Stratton Strawless
NORFOLK
NR10 5NX
01603 755105
wiredbrain@btinternet.com

and other organisations.
- Liaison with UK Government Departments dealing with beekeeping, medicines, and allied matters.
- Liaison and co-operation with UK Beekeeping organisations.
- Contact with European beekeeping organisations (EPBA) and representation on the EEC Honey Working Party (COPA/ COGECA) in Brussels.
- Political lobbying through MPs and Euro MPs.
- Member of the Confederation of National Beekeeping Associations (CONBA)
- Associate member of the Honey Association

FACILITIES FOR MEMBERS:
- Bi-monthly Bulletins with news and updates, notes on meetings with DEFRA, Fera, VMD, and the EEC, reports on current beekeeping problems (e.g. varroa) and commercial developments world-wide.. This bulletin is available as a paper and/or an e-document
  * e-news. Frequent electronic updates on news items
- Free advertisement of members' sales and wants (including hive products, bee stocks and spare equipment).
  * Regional meetings which provide for local discussion and opportunities for trading between members.
- Crop and winter loss reports.
- Free Circulation among members of UK and foreign magazines.
- Free insurance for products and third party liability (not limited to thirty hives).
- Special rates for employers liability insurance.
- Comprehensive special beefarmers insurance with the NFU.
- Pollination contracts.
- Advice from experienced members on all aspects of honey farming and commercial beekeeping; sources of equipment and sundries.
- Product directory listing specialist suppliers.
- Discounts from suppliers.
- Bulk purchase schemes to minimize costs to individual members..

ANNUAL CONVENTION WEEKEND
- Spring meeting for members and partners, held each March at different locations in the UK or abroad. Visits to local bee and research establishments; lectures and discussions on bee-related matters; sight seeing, and social events.

118

GIANT
BEEKEEPING SALE

Only £5 per adult, family tickets available, children go free

Lectures from **DEFRA** on disease management, etc
Over thirty traders including NBB, Thornes, Maisemore, Fragile Planet, Paynes, Swienty, BBWear, Sherriff, ModernBeekeeping,etc

Come and grab an early bargain, see what's new in the world of beekeeping,

Saturday 1st March 2014
9am to 4:30pm

Warwick and Stareton Hall
Stoneleigh Park
Warwickshire
CV8 2LG

www.beetradex.co.uk
info@beetradex.co.uk

Or write
Beetradex Ltd
Unit 5 Maesbury Road Industrial Estate
Oswestry
Shropshire
SY10 8HA

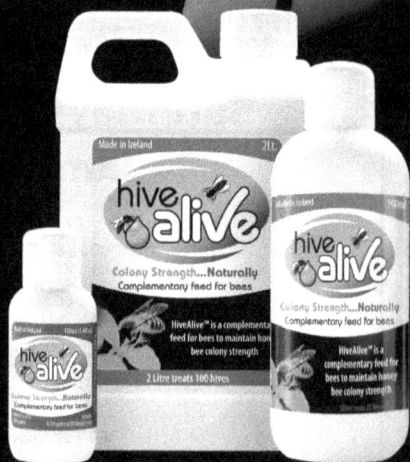

BEEKEEPING EDITORS' EXCHANGE SCHEME

BEES is a self-help grouping of local, county and country beekeeping association editors, which operates principally by exchanging journals through a central address. The scheme is supported by Northern Bee Books.

BEES was founded in 1984 and for many years has been an exchange of paper copy. However, the focus has now changed to an electronic exchange, using the server of one of the participating editors.

Now fully established as part of the British and Irish beekeeping scene, the scheme brings up to date information to beekeepers throughout the British Isles.

B.E.E.S
Helping Editors
Help Themselves

Sponsored by
NORTHERN BEE BOOKS

The aims are:
- to exchange ideas for content and production methods
- to aid others by experience
- to communicate matters editorial
- to share information on national beekeeping issues
- to help and reassure those new to the task
- to give a wider readership to the best writing in beekeeping journalism

If you are an editor or potential editor and would like to know more about how we operate write to Chris Jackson
22 Chapter Close, Oakwood, Derby, DE21 2BG
editors-owner@ebees.org.uk

BEES ABROAD UK Ltd

Supporting beekeeping projects overseas

ADMINISTATOR:
MRS VERONICA BROWN,
Po Box 2058,
Thornbury
Bristol
BS35 9AF.
01173 230 0231;
Info@Beesabroad.Org.Uk

Bees Abroad is a UK-registered charity (No 1108464) established in 1999. Its principle aim is the relief of poverty in the developing world using beekeeping and associated skills as a tool of individual, group and community empowerment for poverty alleviation and to provide sustainable income. Beekeeping is a valuable tool as it is socially and culturally acceptable for both genders across a wide age range. It can cost very little to set up a beekeeping operation, which will deliver benefits for income, education, health, environment and community. Beekeeping and its associated skills deliver access to gainful self-employment for poor and disadvantaged groups. This enables them to recover social status, improve social interactions, obtain income and acquire new skills to build the confidence to represent their own interests. Bees Abroad receives a high volume of direct appeals for assistance from groups all over the world. In practice, it achieves its aims through a volunteer network of supporters, committee members and project managers. Bees Abroad takes care to ensure that its projects are sustainable and not dependent on constant external input. This is done by supporting community group initiatives, setting up village-based field extension services, running training courses for beekeeping trainers and financing local trainers' wages. All Bees Abroad projects are designed to become self-financing after a defined time period. Its first two projects in Nepal and Cameroon now employ 42 beekeeper trainers and

involve many more. It currently has projects either running or seeking funding in Malawi, Kenya, Ghana, Tanzania, Uganda and Nigeria.

Bees Abroad is run by volunteers, who are all beekeepers. They currently undertake all activities, including fundraising, though an administrator is employed for one day a week. We also arrange beekeeping holidays to variety of locations, including Morocco, Poland and Nepal.

For more details of what we do and how you can help, contact Veronica Brown, the Administrator, Bees Abroad. You can learn more about our work and make a regular or on-off donation through our website, www.beesabroad.org.uk

BEES *for* DEVELOPMENT TRUST

supporting beekeepers in developing countries

www.beesfordevelopment.org

We always need more help and skills –
please contact us if you might like to be involved.

Bees *for* Development
1 Agincourt Street
Monmouth
NP25 3DZ
info@beesfordevelopment.org
01600 714848
UK registered Charity
No 1078803

We are a professional organisation, working in the beekeeping development sector for 20 years. We are respected and trusted by beekeepers world-wide.

Worldwide

- Providing training and information materials for community groups to improve their knowledge of beekeeping and business
- Publishing **Bees *for* Development Journal** keeping remote beekeepers in 130 countries up to date with news, practical advice, and events
- Maintaining a large resource base for the sector, available on-line

International Projects

With specific communities and partners we work to increase beekeeping incomes

- Cameroon – better processing equipment to raise quality of saleable produce
- Ethiopia – strengthening market chains and training trainers
- Kyrgyz Republic – helping to solve land use conflicts for beekeepers
- Uganda – enabling beekeepers to access good markets

Shop in Monmouth

All proceeds from sales go to support our charitable work. We sell

- A wide range of local honeys and bee products, African honey
- Beekeeping equipment
- bee-related gifts, cards and books

Information Gallery in Monmouth

- Learn about our work overseas
- See a range of unusual bee hives
- Find out about courses and events

YOU CAN HELP US BY:

- **Sponsoring** a Journal for a beekeeper working in a poor country
- **Making** a gift of a Resource Box for a training in a school or project
- **Giving** a donation
- **Joining** one of our *Beekeepers' Safaris*
- **Buying** from our shop in Monmouth or from our on-line store
- **Attending** one of our *Courses*
- **Offering** your skills to work with us as a volunteer
- **Ensuring** that your group or organisation knows about our work, and supports us if possible
- **Helping** us to present our work at events

BRITISH BEEKEEPERS'
ASSOCIATION www.bbka.org.uk

COMMITTEES OF THE EXECUTIVE AND SECRETARIES

FINANCE
This team of Trustees reviews & agrees all budgets, handles all investment matters, finalising insurance policies and sets proposals relating to capitation.

Governance
Primary areas of responsibility are to ensure that we adhere to Charity Commission rules, that we operate within the constitution in addition to ensuring that our Trustees act in the best interests of the BBKA and it's members.

Operations & Membership Services
Headed by the Vice Chairman this team ensure that all Membership Services are administered effectively and on time and that the organisation operates efficiently. It also acts as a co-ordinator for all external fundraising.

Technical & Environmental
All technical issues and their their potential impact on bees and beekeeping are assessed and monitored within this team. All research projects are reviewed and recommendations made by Technical & Environmental group.

Public Affairs
Whether it be government liason, both UK & EU, or press activity this comes from the Public Affairs team.

OPERATIONS DIRECTOR & GENERAL SECRETARY
Jane Moseley
The British Beekeepers' Association,
National Beekeeping Centre,
Stoneleigh Park
Warwickshire CV8 2LG
024 7669 6679
Fax: 024 7669 0682
Email:
jane.moseley@bbka.org.uk

BRITISH BEEKEEPERS ASSOCIATION
National Beekeeping Centre
Stoneleigh Park,
Kenilworth, Warks CV8 2LG
02476 696679
Fax: 024 7669 0682
Office hours 9.00am–5.00 pm
Monday - Friday (inclusive)
Telephone answering service outside office hours
Email:
bbka.info@bbka.org.uk
Web: www.bbka.org.uk

BBKA

✉ ☎

PRESIDENT
MARTIN TOVEY
martin.tovey@bbka.org.uk

CHAIRMAN
DR DAVID ASTON
david.aston@bbka.org.uk

VICE CHAIRMAN
DOUG BROWN
doug.brown@bbka.org.uk

TREASURER
Shena Winning
treasurer@bbka.org.uk

All enquiries should be made to BBKA Press Officer: gill.maclean@bbka.org.uk

Education & Training

The development of information from practical guidance notes, advisory leaflets, training materials while also undertaking it's own educational initiatives in support of improving the knowledge and skills of beekeepers at all levels. Education & Training liase with the Examination Board to develop training materials to support Association tutors with products such as the Course in a Case.

Examination Board

The BBKA examination board provide a structured range of examinations fulfilling the needs of all beekeepers from Junior Certificate to Master Beekeeper. The board are responsible for all matters relating to the syllabus, content and assessment and operate independently of the BBKA board of Trustees. Where Associations have no Examinations Secretary the Association Secretary deals with examinations. To help future candidates it is suggested that Associations without an Examination Secretary appoint one. Associations are responsible for arranging a suitable room for the written examinations and recommending an invigilator.
Contact Val Frances, Exam Board Secretary
Email: val.frances@bbka.org.uk Tel 01226 286341

Insurance

Members of BBKA, Area Associations and officials are indemnified against claims for Public Liability to a limit of £10million, Product Liability to a limit of £10 million, Professional Indemnity to a limit of £2 million relating to their beekeeping activities. BBKA Association Officer and Trustee liability insurance also applies to a limit of £10 million. Each new claim carries an excess payable by the member.
An 'All Risks' policy is available to both individuals and Associations, to cover the loss or damage of property & equipment. Details are available via www.bbka.org.uk or the main office: 02476 696679

Publications
- BBKA News is issued monthly free to all members of the BBKA, featuring articles about bees, beekeeping and the other associated articles of interest. Editorial: editorial@bbkanew.org.uk Advertising: advertising@bbkanews.org.uk
- BBKA Year Book is published each Spring and is for Association use and reference.It contains detailed information on the BBKA including useful reference tools such as a directory of Lecturers and Demonstrators.
- Members Handbook is published annually and sent to new members
- BBKA Introduction to Beekeeping

BBKA Website - www.bbka.org.uk
The BBKA Website contains technical information, is easy to navigate and supports both beekeepers and the general public. You can download publications, find help and advice in the discussion forums, purchase merchandise, learn about Bees, use the Bees4kids section, download BBKA exam application forms and the exam syllabus. Within the Members Only area, specific insurance downloads and other member only information is available. Associations beekeeping events are promoted.
A Swarm Collector database is included within the site enabling the general public with a direct link to a local swarm collector.

Events
Area and local associations attend and exhibit at various events within their local throughout the year while the BBKA supports selected national shows. Whether it be village fete or national exhibition these events continue to provide a vital service for the dissemination of knowledge.

BBKA Spring Convention
Held in April every year this is a firmly established major beekeeping event. Lectures and Workshops are

staged over 3 days with a one day trade exhibition on the Saturday. Both Friday and Sunday are member only days which are ticketed.

Slide Library
The BBKA slide library has been digitised for ease of use and preservation. For a list of slides available and their format please go the BBKA Members Area at www.bbka.org.uk or contact the BBKA office.

Subscriptions & Membership Fees
Individual Membership of the BBKA is £37 per annum, for an Overseas Member the fee is £28.00. All other membership is via local associations.
Friends Membership is also available via www.bbka.org.uk

Exam Board Footnote
Where Associations have no Examinations Secretary the Association Secretary deals with examinations. To help future candidates it is suggested that Associations without an Examination Secretary appoint one. Associations are responsible for arranging a suitable room for the written examinations and recommending an invigilator.
If you live in an area without a nominated Exam Secretary, you should contact Val Frances, Exam Board Secretary Email: val.frances@bbka.org.uk Tel 01226 286341

BBKA Enterprises
BBKA Enterprises Ltd is a private company, limited by guarantee with all profits from the trading activities being donated to the BBKA. Via the BBKA online shop a range of beekeeping, corporate and related items, specially selected books, gifts, travel items and educational materials are available.
Visit www.thepollenbasket.com, the official BBKA web shop, or call 02476 696679

BBKA

✉ ☎

AREA ASSOCIATION SECRETARIES

AVON, Rosemary Taylor
43 High Street
Chew Magna
Bristol BS40 8PR
01275 332 438
rosemary.taylor@tiscali.co.uk
BERKSHIRE, Martin Moore,
19 Armour Hill
Tilehurst
READING
RG31 6JP
01189677386
07729620286
secretary.berksbees@
uwclub.net
BOURNEMOUTH
Peter Darley
3 Dorset House
Branksome Park
Poole BH13 6HE
01202 767654
07825 014757
peter.darley@talktalk.net
BUCKS, Sue Carter
sue.carter@kodak.com
CAMBRIDGESHIRE
Susanna Fenwick
27 Pratt Street
Soham
Cambridgeshire CB7 5BH
07837 977399
susfen@hotmail.com
CHESHIRE, M.F. Haynes
98, Gatley Road, Gatley,
Cheadle,
Cheshire SK8 4AB
0161 491 2382
thesecretary@cheshire-bka.
co.uk

CHESTERFIELD, Robin Bagnall
21 Ramper Avenue,
Clowne, Chesterfield
Derbyshire S43 4UD
01246 570545
ancient.mariner74-79@
virgin.net
CLEVELAND, Derek Herring
8 Wardale Avenue
Middlesborough TS5 8TH
01642 282030
derek.herring@ntlworld.com
CORNWALL, Julia Cooper
Whistow Farm, Lanlivery,
Bodmin, PL30 5DE
01208 872865
julia.i.cooper@btinternet.com
CORNWALL WEST
Dr B Doeser
Caerlaverock
Tresowes Hill
Ashton
Helston TR13 9TB
01736 763876
secretary@westcornwallbka.
org.uk
CUMBRIA, Stephen C Barnes
8 Albemarle Street
Cockermouth
Cumbria CA13 0BG
01900 824872
braithwaitebees@sky.com
DERBYSHIRE, M J Cross
Harlestone, Beggarswell
Wood, Ambergate
Derbyshire DE56 2HF
01773 852772
crosssk@btinternet.com
DEVON, Colin Sherwood
01404 42130

c.j.sherwood@btinternet.com
DORSET, Mrs Ruth Homer
5, Malters Cottage,
Litton Cheney,
Dorchester DT2 9AE
beekeepers@hotmail.com
DOVER & DISTRICT
Mrs Maggie Harrowell
4 Harton Cottages, Ashley,
Dover, CT15 5HS
01304 821208
the.harrowells@btinternet.com
DURHAM, Bala Nair
39 Windsor Drive
Catchgate
Stanley
Co. Durham DH9 8SR
01207 236 839
lawrencian66@hotmail.com
ESSEX, Mrs Pat Allen
8 Frank's Cottages
St Mary's Lane
Upminster RM14 3NU
01708 220897
pat.allen@btconnect.com
GLOUCESTERSHIRE
Marie Toman
Oak Cottage,
Stoulgrove Lane, Woodcroft,
Chepstow, Mon., NP16 7QE.
01291 620345
marietoman@btconnect.com
GWENT, William Stewart
01600 740665
secretary@gwentbeekeep-
ers.co.uk
HAMPSHIRE, Mrs P Barker
Brookdean, Hillbrow
Liss, Hampshire GU33 7PT
01730 895368

BBKA

✉ ☎

H'GATE & RIPON
Peter Ross
The Wheelhouse
The Green Scriven
Knaresborough HG5 9EA
01423 866565
secretary@hrbka.org.uk
HEREFORDSHIRE
Anne Harvey
14 Burghley Close
Stevenage SG2 8SX
01438 361584
secretary@hertsbees.org.uk
HERTFORDSHIRE
Luke Adams
53 Park Street Lane
Park Street
St. Albans
Hertfordshire AL2 2JA
01442 843 779
luke.skywalker@virgin.net
HUNTINGDON,
Nick Steiger
Bull Cottage, Main Street,
Upton, Huntingdon, Cambs,
PE28 5YB
01480 891935
n.steiger@btinternet.com
INSTITUTE OF NORTHERN IRE-
LAND BEEKEEPERS
Caroline Thomson
105 Cidercourt Road
Crumlin BT29 4RX
0289 445 3655
secretary@inibeekeepers.
com
ISLE OF MAN,
Janet Thompson
Cott ny Greiney, The Smelt,
Beach Road, Port St Mary,
Isle of Man, IM9 5NF
01624 835524
jthompson@manx.net

ISLE OF WHITE,
Mrs. Mary Case
Limerstone Farm
Limerstone,
Newport
Isle of White PO404AB
01983 759510
gcase90337@aol.com
KENDAL &
SOUTH WESTMORLAND
Peter Llewellyn
1 Greenside House
Hincaster
Milnthorpe LA7 7NA
015395 62369
pdwllewellyn@yahoo.co.uk
KENT, Mrs J D Spon-Smith
77 Bushey Way
Beckenham
Kent
020 8663 1364
jennifer@spon-smith.com
LANCASHIRE & NORTH WEST
Martin Smith
137 Blaguegate Lane,
Lathom
Skelmersdale
Wigan WN8 8TX
07831 695732
ormskirk_beekeepers@
hotmail.com
LANCASTER
Derek Wright
Bridge Barn
Tatham
Lancaster LA2 8NL
01524 222577
dertine@hotmail.com

LINCOLNSHIRE
Mrs. Celia Smith
Brookfield, Moor Town Road,
Nettleton LN7 6HX
07527 600698
sec.lincsbka@yahoo.com
LONDON, Angela Woods
4 The Carlton Tavern
73 Grafton Road
London NW5 4BB
07850263077
sec@lbka.org.uk
LUDLOW & DIST
Trisha Marlow
Willow Farm
Old Church Stoke
Montgomery SY15 6DQ
07812 518 822
susbees@gmail.com
MANCH. & DIST, Mrs. M. Bohme
54 Dunster Drive, Flixton
Manchester M41 6WR
0161 747 7292
MEDWAY, Sheila Stunell
01634 270 961
07718985910
sheila.stunell1@btinternet.
com
MIDDLESEX, Mary Hunter
020 8367 8452
mary@hunter67.myzen.co.uk
NEWBURY & DISTRICT
Virginia Arnott
43 Hendred Way
Abingdon
Oxon OX14 2AW
v.arnott@ntlworld.com
NEWCASTLE & DISTRICT,
Brian Diver
12, Coquet Gardens
Wallsend
Tyne & Wear
NE28 6AG
bdiver@accesstraining.org

NORFOLK, Louise Hutchinson
11 Glebe Crescent
Cawston
Norwich NR10 4BD
01603 872042
www.norfolkbeekeepers.
co.uk/contact/contact_the_
secretary.php
NORFOLK WEST & KINGS LYNN
Kay Marshall
123 Main Road
Quadring
Spalding
Lincs PE11 4PJ
01775 821478
sec@wnklba.co.uk
NORTHAMPTONSHIRE
Mrs Ruth Stewart
17 Leys Avenue, Rothwell,
Kettering,
Northants, NN14 6JF
01536 507293
rstewart@euramax.co.uk
NORTHUMBERLAND
Revd Benjamin Hopkinson
benjaminhopkinson@
hotmail.com
NOTTINGHAMSHIRE
M. Jordan
29 Crow Park Avenue
Sutton on Trent
Nr Newark NG23 6QG
01636 821613
mauricejordan11@btinternet.
com
OXFORDSHIRE,
Roya Haghighat-Khah
obka.sec@gmail.com
PETERBOROUGH & DISTRICT
P George Newton
65 Queen Street, Yaxley
Peterborough PE7 7JE
01733 243349

RUTLAND
Fliss Haynes
16 Shannon Way
Oakham LE15 6SY
fliss@rbka.org
SEDBURGH
Jane Callus-Whitton
Harren House, Woodman
Lane, Cowan Bridge,
Carnforth, LA6 2HT
01524 272004
janeecwhitton@yahoo.co.uk
SHROPSHIRE, Chris Currier
Churchleigh
Adderley
Market Drayton TF9 3TD
01630 655 422
chris.currier@btopenworld.
com
SHROPSHIRE NORTH
Nigel Hine
Chapel House
Post Office Lane
Whixall
Shropshire SY13 2RL
01948 880052
nahine@btinternet.com
SOMERSET,
Dr Richard Bache
secretary@
somersetbeekeepers.org.uk
STAFFORDSHIRE NORTH
David Teasdale
Bosworth House
Station Road
Endon
Stoke on Trent ST9 9DR
01782 502 495
secretary@northstaffsbees.
org.uk

STAFFORDSHIRE SOUTH
Mrs Lynne Lacey
21 Fisherwick Road
Whittington
WS14 9LL
01543 432202
lynne.lacey123@
btinternet.com
STRATFORD-ON-AVON
Michael Osborne
Oak Lodge, King's Lane
Snitterfield
Stratford-upon-Avon
Warwickshire CV37 0RB
01789 731745
mjroosborne@btinternet.com
SUFFOLK,
Ian McQueen
643 Foxhall Road
Ipswich IP3 8NE
01473 420187
jackie.mcqueen@ntlworld.com
SURREY, Mrs Sandra Rickwood
19 Kenwood Drive,
Walton-on-Thames
Surrey. KT12 5AU
01932 244 326
rickwoodsbka@googlemail.com
SUSSEX, Liz Twyford
Westcott
Udimore Road
Broad Oak
Rye TN31 6DG
01424 882361
secretary@sussexbee.org.uk
SUSSEX WEST,
Mr John Glover
Graham Elliott
Robins Croft
Chalk Road
Ifold
Loxwood
Billingshurst RH14 0UB
grahammt@tiscali.co.uk

BBKA

✉ ☎

THANET,
Mrs R Pearce
Summerfield Cottage
Summerfield
Woodnesborough
Nr Sandwich CT13 0EW
01304 614789

VALE & DOWNLAND
Mrs Jane Greenhalgh
8 Park View
Garston Lane
Wantage
Oxon OX12 7AQ
01235 760157
jane.greenhalgh@lineone.net

TWICKENHAM & THAMES VALLEY
(MOLE APIARY CLUB)
Mrs Sarah Crofton
11 Wellesley Avenue
London TW3 2PB
0208 222 8216

WARWICKSHIRE,
Nigel Fleming
14 Underwood Road
Handsworth Wood
Birmingham B20 1JH
0121 240 0263
nigel.fleming@blueyonder.co.uk

WILTSHIRE,
Ruth Woodhouse
Sandridge Tower
Bromham
Devizes
SN15 2JN
01225 705382
sandridgetower@aol.com

WORCESTERSHIRE
Mr Chris Broad,
Upper Gambolds Farm,
Upper Gambolds Lane,
Bromsgrove
Worcestershire
B60 3HD

01527 872448
chrisbroad1964@btinternet.com

WYE VALLEY, Mrs S Wenczek
Susan Quigley
New House farm
Michaelchurch Escley
HR2 0PT
019815 10183

YORKSHIRE, Brian Latham
111 Woodland Road
Whitkirk, Leeds
LS15 7DN
0113 264 3436
chrisbroad1964@btinternet.com

ASSOCIATION EXAMINATION SECRETARIES

AVON, Position Vacant
Please contact
Hon. General Secretary
Julie Young
01179 372 156

BERKSHIRE,
Mrs Rosemary Bayliss
Norbury, Coppid Beech Hill,
Binfield, Berkshire.
RG42 4BS
01344 421747

BOURNEMOUTH, Mrs. M. Davies
57 Leybourne Avenue
Ensbury Park
Bournemouth
Dorset BH10 6ES
01202 526077

BUCKS, John Chudley
Orchard Lea, Oxford Street
Lea Common
Great Missenden HP16 9JT
01494 837544
jlchudley@tiscali.co.uk

CHESHIRE
Graham Royle NDB,
7, Symondley Road,
Sutton,
Macclesfield. SK11 0HT
01260 252 042

CORNWALL
Mrs. Susan Malcolm
Fig Tree, 333 New Road
Saltash, Cornwall
PL12 6HL

01752 845496

DEVON, Roger Lacey
Gatchell House
Toadpit Lane, Ottery St Mary
Devon EX11 1TR
01404 811733
devonbees@pobox.com

DORSET, K.G.Bishop
72 Alexandra Road
Bridport DT6 5AL
01308 425479

DURHAM, G. Eames
23 Lancashire Drive
Belmont, Durham,
DH1 2DE
01913 845220
george.eames@durham.ac.uk

ESSEX, Pat Allan
8, Frank's Cottages
St. Mary's Lane
Upminster, RM14 3NU
pat.allen@btconnect.com

GLOUCESTERSHIRE
Bernard Danvers
120a Ruspidge Road
Cinderford,
Gloucestershire
GL143AG
01594 825063

GWENT, Mrs J Bromley
Ty Hir, Monmouth Road
Raglan, Usk. NP15 2ET
01291 690331
bromleyjan@hotmail.com

HAMPSHIRE, Mrs Peggy Mason
37 Springford Crescent
Lordswood,
SO16 5LF
023 8077 7705

H'GATE & RIPON, Peter Ross
The Wheelhouse, The Green,
Old Scriven, Knaresborough
HG5 9EA, 01423 866565,
pjeross@btinternet.com

HEREFORDSHIRE, Len J. Dixon
The Square, Titley,
Kington
Herefordshire HR5 3RG
01544 230884
beeline2ljd@yahoo.co.uk

HERTFORDSHIRE, R. E. A. Dart-ington
15 Benslow Lane
Hitchin SG4 9RE
01462 450707
gray.dartington@dial.pipex.com

ISLE OF WIGHT, Mrs M. Case
Limerstone Farm,
Limerstone, Newport,
Isle of Wight, PO30 4AB
01983 740223
gcase90337@aol.com

KENT, P. F. W. Hutton
22 Good Station Road
Tunbridge Wells,
TN1 2DB
01892 530688

LANCASHIRE & NW
Edward Hill
3 Sandy Lane, Aughton
Ormskirk
L39 6SL
01695 423137

LEICESTERSHIRE & RUTLAND
Brian Cramp
2 Woodland Drive, Groby
Leicester
LE6 0BQ
01162 876879

LINCOLNSHIRE, R. J. B. Hickling
Linden Lea, Sandbraes Lane,
Caistor, LN7 6SB
01472 851473

MIDDLESEX
Mrs Jo V Telfer
Midwood House
Elm Park Road, Pinner
Middlesex HA5 3LH
020 8868 3494
e-mail, jvtelfer@hotmail.com

NOTTINGHAMSHIRE
Dr Glyn D Flowerdew
Knight Cross Cottage
Newstead Abbey Park
Ravenshead
Nottinghamshire NG15 8GE
01623 792812

OXFORDSHIRE, Terry. Thomas
4 Kirk Close
Oxford, OX2 8JN
01865 558679

PETERBOROUGH, P. G .Newton
65 Queen Street, Yaxley
Peterborough PE7 3JE 01733
243349

SHROPSHIRE NORTH
Paul Curtis
1 Hammer Close
Overton-on-Dee, Wrexham
Clwyd LL13 0LD0
01691 624296

SOMERSET, Mrs Angela Bache
Greenway House
Badgers Cross
Somerton TA11 7JB
Tel 01458 273149

STAFFS.Nth Dr. Nick C Mawby
Glenwood, Wood Lane
Longsdon,
Stoke on Trent ST9 9QB
01538 387506
info@northstaffsbees.org.uk

STAFFS. SOUTH
Tony Burton
96 Weeping Cross, Stafford,
Staffordshire. ST9 9QB
01538 399322

SUFFOLK, Mr Ian McQueen
643 Foxhall Road, Ipswich,
Suffolk, IP3 8NE
01473 420187

SURREY, Mrs. A. Gill
143 Smallfield Road
Horley, RH6 9LR
01293 784161

BBKA

✉ ☎

SUSSEX, Nigel Champion
45 Ridgeway,
Hurst Green
Etchingham
East Sussex TN19 7PJ
01580 860379
SUSSEX WEST
Mrs A. S. Gibson-Poole
Mont Dore, West Hill
High Salvington
Worthing, BN13 3BZ
01903 260914
TWICKENHAM, Chris Deaves
12 Chatsworth Crescent
Hounslow,
Middlesex
TW3 2PB
0208 5682869
e-mail,
c-deavs@compuserve.com

WARWICKSHIRE, P.D. Lishman
Aston Farm House
Newtown Lane
Shustoke, ColeshillB46 2SD
01676 540411
WILTSHIRE, John Troke
The Lythe
Hop Gardens
Whiteparish, Salisbury,
Wiltshire SP5 2SS
01373 822892

WORCESTERSHIRE, D.P. Friel
17 Tennal Rd, Harborne
Birmingham, B32 2JD
0121 427 1211
YORKSHIRE, Brian Latham
Tel: 0113 264 3436
Mob: 07765842766
brian.latham@ntlworld.com

Where Associations have no Examinations Secretary the Association Secretary deals with examinations. To help future candidates it is suggested that Associations without an Examination Secretary appoint one. Associations are responsible for arranging a suitable room for the written examinations and recommending an invigilator.

If you live in an area without a nominated Exam Secretary, you should contact Mrs Val Frances, 39 Beevor Lane, Gawber, Barnsley, S75 2RP Tel 01226 286341. e-mail, valfrances@blueyonder.co.uk

HOLDERS OF THE BBKA SENIOR JUDGES CERTIFICATE

ASHLEY, Mr. T. E.
Meadow Cottage
Elton Lane, Winterley
Sandbach
Cheshire CW11 4TN

BADGER, M.J , MBE
14 Thorn Lane
Leeds, LS0 1NN

BLACKBURN, Mrs. H.M
15 Highdown Hill Road
Emmer Green
Reading RG4 8QR

BROWN, Mrs. V

BUCKLE, M.J
The Little House
Newton Blossomville
Bedford MK43 8AS
01234 881262
martin@newtonbee.fsnet.
co.uk

CAPENER, Rev. H.F.
1 Baldric Road
Folkestone CT20 2NR

COLLINS, G.M. , NDB
72 Tatenhill Gardens
Doncaster DN4 6TL

COOPER, Miss R.M
10 Gaskells End
Tokers Green
Reading RG4 9EW

DAVIES, Mrs. M
57 Leybourne Avenue
Ensbury Park
Bournemouth BH10 6HE

DIAPER, B
B Diaper
57 Marfield Close
Walmley
West Midlands
0121 313 3112
or 07711 456932

DICKSON, Ms. F
Didlington Manor
Didlington, Thetford
Norfolk IP26 5AT

DUFFIN, J.M
Upper Hurst
Salisbury Road, Blashford
Ringwood
Hampshire BH24 3PB
01425 474552

FIELDING, L.G
Linley, Station Road
Lichfield WS13 6HZ

MacGIOLLA CODA, M.C.
Glengarra Wood, Burncourt
Cahir, Co. Tipperary
Republic of Ireland

McCORMICK, E.
14 Akers Lane, Eccleston St.
Helens, Lancs WA10 4QL

MOXON, G
9 Savery Street
Southcoates Lane
Hull HU9 3BG

ORTON J
Occupation Road, Sibson
Nuneaton CV13 6LD

SALTER T.A , MBE
44 Edward Road, Clevedon
North Somerset BS21 7DT

SYMES, C.J
189 Marlow Bottom Road
Marlow SL7 3PL

TAYLOR, A.J
The Old Pyke Cottage
Hethelpit Cross, Staunton
GL19 3QJ

WILLIAMS, M
Tincurry, Cahir,
Co Tipperary, Eire

YOUNG, M
Mileaway, Carnreagh Hills-
borough,
Northern Island BT26 6LJ

BIBBA

BEE IMPROVEMENT & BEE BREEDERS' ASSOCIATION

<u>www.bibba.com</u>

SECRETARY
Roger Cullum-Kenyon
Craig Fawr Lodge
Caerphilly
CF83 1NF.
02920 869242
dinah@dinahsweet.com

MEMBERSHIP SECRETARY
Iain Harley
93 Dunsberry
Bretton
Peterborough
Cambridgeshire
PE3 8LB
01733 700740
iain.harley42@ntlworld.com

SALES SECRETARY
John Hendrie
26 Coldharbour Lane
Hildenborough
Tonbridge
Kent
TN11 9JT
sales@bibba.com

BIBBA is an organisation devoted to encouraging beekeepers to breed native bees. The bee more suited to our environmental circumstances than other sub species. BIBBA's aims are publicised through books, workshops, lectures and conferences.

BIBBA also co-operates with worldwide Beekeeping and breeding groups interested in conserving and improving their own native bees.

Breeding techniques advocated include:

- Assessment of colonies by observation, recording certain criteria on standard record cards.
- Determination and purity of sub species by measurement of morphometric characters and mitrochondial DNA.
- Use of mini nucs for the mating of queens economically

BIBBA Publications include:

- The Honeybees of the British Isles by Beowulf Cooper
- Breeding Techniques and Selection for Breeding of the Honeybee by Prof. F. Ruttner
- The Dark European Honey Bee by Prof. F. Ruttner, Rev. Eric Milner and John Dews
- Breeding Better Bees using Simple Modern Methods by John E. Dews and Rev.Eric Milner
- Better Beginnings for Beekeepers by Adrian Waring - second edition.

BIBBA encourages the formation of Bee Breeding Groups, and the sharing of knowledge between groups by the provision of genetic material.

Look out for Queen Rearing events in the bee press and on www.bibba.com.

CABK

✉ ☎

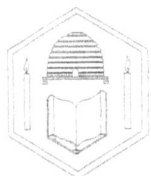

THE CENTRAL ASSOCIATION OF BEEKEEPERS

www.cabk.org.uk

SECRETARY, Pat Allen
8 Frank's Cottages
St Mary's Lane
Upminster, RM14 3NU
pat.allen@btconnect.com

PRESIDENT, Prof. R.S. Pickard
Consumer's Association
pickard.r@btopenworld.com

TREASURER, John Hendrie
26 Coldharbour Lane
Hildeborough
Tonbridge, TN11 9JT
bibba26@talktalk.net

PROGRAMME SECRETARY
Pam Hunter
Burnthouse
Burnthouse Lane
Cowfold, Horsham
RH13 8DH
pamhunter@burnthouse.org.uk

EDITOR, Pat Allen
8, Frank's Cottages
St. Mary's Lane
Upminster, RM14 3NU

SALES AND DISTRIBUTION,
Margaret Thomas
Tighnabruaich,
Taybridge Terrace,
Aberfeldy, Perthshire
PH15 2BS
zyzythomas@waitrose.com
01887 829710

The Central Association of Beekeepers in its present form dates from the time of the reorganisation of the British Beekeepers' Association in 1945. The BBKA was originally made up of private members only. However as County Associations were formed they applied for affiliation and were later permitted to send delegates to meetings of the Central Association, as the private members were then known. This arrangement became unsatisfactory as the voting power of the Central Association greatly outnumbered that of the County Associations and so in 1945 a new Constitution was drawn up whereby the Council comprised Delegates from the Counties and Specialist Member Associations. The private members then formed themselves into a Specialist Member Association with the designation 'The Central Association of the British Beekeepers' Association'; this was later shortened to its present style.

The Association was able to devote itself to its own particular aims, to promote interest in current thought and findings about beekeeping and aspects of entomology related to honey-bees and other social insects. Lectures given by scientists and other specialists are arranged, printed and circulated to members, as has been done since 1879.

An annual Spring Conference is held in London and an Autumn Conference in the Midlands. In addition, a lecture is presented at the Annual General Meeting and at the Social Evening held during the National Honey Show. The subscription is £10.00 per annum, £12.00 for dual membership (one copy only of publications).

BEEKEEPING MAGAZINES AND NEWSLETTERS

Australasian Beekeeper
Editor, Des Cannon
Pender Beegoods P/L
PO Box 7124
Karabar NSW 2620
Australia
+61 2 6236 3294
editor@theabk.com.au
www.thweabk.com.au
Subscriptions:
International: $160.00AUD
Published monthly,
12 Issues/year

Bedfordshire BKA
Editor, Sue Lang
154a Lower Shelton Rd
Upper Shelton
Marston Moretaine
Beds. MK43 0LS
01234 764180
07879 848550
bedfordshirehoney@
hotmail.co.uk
www.bedfordshirebeekeepers.
org.uk

Bee Craft Bee
Editor, Claire Waring
editor@bee-craft.com
£27 per year

Bee Culture
Editor, Kim Flottum
one year subscription surface
mail is $47.50; two years
$89.00 and three years
$130.00.
An airmail one-year
subscription is $100.00.

Bee World
£33 per year
mail@ibra.org.uk
Journal of Apicultural
Research
£82.50 per year
mail@ibra.org.uk

Beekeeper's Quarterly, The
Editor, John Phipps
manifest@runbox.com
The Companion to the
Beekeepers Annual
Subscriptions £30 p.a
(but group schemes at
reduced rates exist for BKAs)
from: Northern Bee Books

Berkshire Federation
Newsletter
Editor, Sue Nugus
sue@
academic-conferences.org

Blackburn &
East Lancashire
Editor, Michael Birt
webmaster@
blackburnbeekeepers.com

Bournemouth & South Dorset
Monthly newsletter
Editor, Peter Darley,
3 Dorset House
42 The Avenue
Branksome Park
Poole, Dorset.
BH13 6HE
07825 014757
Peter.Darley@talktalk.net

Bradford Newsletter
editor, Bill Cadmore
http://www.bfdbka.org.uk
bill.cadmore@ntlworld.com
W P Cadmore
Horsforth Blossom Honey
104 Hall Lane, Leeds
LS18 5JG
0113 216 0482
07847 476927

BEE MAGS

Cambridge Newsletter
Editor, Paul Schofield
jpaul.schofield@btinternet.com

Cheshire Beekeeper
Editor, Lesley Jacques
39 The Crescent
Northwich
CW9 8AD
07788 744086 / 01606 333272

Cornwall
Editor, Gillian Searle
gwenynkemow@tiscali.co.uk

Cumbria Bee Times
Editor, Val Sullivan
Brackenwray Farm,
Kinniside, Cleator,
Cumbria,
CA23 3AG
01946 862604
brackenwray@aol.com

Derbyshire BKA
Editor, Jim Parish
jamesparish382@btinternet.com
01332 781509

Devon Beekeepers Magazine
Editor, Mrs Lilah Killock
editor@devonbeekeepers.org.uk

Dorset
Honeycraft
comes out 4 times a year March,
June, September and December
Editor, Lesley Gasson
The White House,
Candy's Lane,
Shilllingstone.
DT11 0SF
01258 861690
lmgasson@btinternet.com

Durham
Editor, George Eames
11, Sharon Avenue,
Kelloe, Durham
DH6 4NE
07970 926250
beeseames@btinternet.com

Essex Beekeepers.
Monthly Magazine
'The Essex Beekeeper'
Editor, Jean Smye
jsmye@o2.co.uk
07731 856361

Gloucestershire
Editor, Annie Ellis
19 Whaddon Road
Cheltenham
Gloucestershire
GL52 5LZ

Harrogate
Editor, Judith Rowbottom
judithrowbottom@hotmail.com

Herefordshire Newsletter
Editor, Lin Steppes
lin@thesteppeshereford.co.uk

Ipswich & East Suffolk
Editor, Jeremy Quinlan
jeremyq@tiscali.co.uk
01473 737700.

Kendal & South Westmorland
Editor, Peter Llewellyn
pdwllewellyn@yahoo.co.uk
01539 562369

Kent
Editor, Jenny Spon-Smith
jennifer@spon-smith.com
020 8663 1364

Leeds BKA
Editor, Kathleen Slater
katey.slater@ntlworld.com

Leicester & Rutland
Editor, Simon Skerritt
1 Orchard Lane,
Countesthorpe,
Leicestershire
LE8 5RB
nudey.hippy@gmail.com

Manchester
MDBKA
liz.sperling@mdbka.com
0161 792 5468

Newcastle BKA
Editor, Armele Philpotts
newcastlebeekeepers@yahoo.
co.uk

New Zealand Bee Journal
Editor, Pauline Downie
ceo@nba.org.nz
Sub NZ$176

Norfolk
Norfolk Beekeepers'
Federation Yearbook 2014
NBKA Newsletter Buzzword
Editor, Michael Lancefield

Nottingham
Editor, Stuart Ching
122 Marshall Hill Drive
Porchester
Notts
NG3 6HW
jsching37@yahoo.co.uk

Somerset Newsletter
Somerton
Editor, Stewart Gould
(editor for both
Somerton & Somerset)
1 The Folly
Ditcheat
Shepton Mallet
Somerset
BA4 6QS
01749 860755
somertonbees@aol.com

Stratford on Avon
Editor, Peter Edwards
Keeper's Cottage
Alne Hills
Great Alne
Alcester
B49 6JU
01789 488020, 07913 709076
www.stratfordbeekeepers.org.uk

The Speedy Bee -
ceased publication

Warwickshire BKA
Editor, Julia Barclay
wbeditor@
warwickshirebeekeepers.org.uk

Welsh BKA Magazine
Editor, Sue Closs
editor@wbka.com

West Cornwall
Published each month
Eeditor, Barbara Barnes
membership@
westcornwallbka.org.uk
07901 977 597

Yorkshire Newsletter
Editor Bill Cadmore -
see Bradford for details

CONBA

CONBA-UK & Ireland
COUNCIL OF NATIONAL BEEKEEPING ASSOCIATIONS IN THE UNITED KINGDOM and IRELAND

Incorporating the beekeeping organisations of :
England, Channel Islands Isle of Man, Wales, Scotland, Ulster, Ireland and The Bee Farmers Association

SECRETARY David Bancalari (BFA)
Park Farm Barn
Shorthorn Road
Stratton Strawless
Norfolk. NR10 5NX
01603 755105
wiredbrain@btinternet.com

CHAIRMAN, Mervyn Eddie (UBKA)
3b Old Road
Upper Ballinderry
Lisburn, Co. Antrim. BT28 2NJ
0289 265 2580
eddie_mervyn@yahoo.co.uk

VICE CHAIR Michael Gleeson (FIBKA)
Ballinakill
Enfield
Co.Meath, Ireland.
+353 (0)4695 41433
mgglee@eircom.net

HON.TREASUER Martin Tovey (BBKA)
11, Coach Road
Warton, Carnforth
Lancashire. LA5 9PP
01524 730451
martintovey@hotmail.co.uk

CONBA was established in 1978 to promote the aims and objectives of the national beekeeping associations of England, Scotland, Ulster, Wales and Ireland, and the Bee Farmers Association. Its purpose is to represent the interests of beekeepers' with local, national and international authorities. A representative delegate from each of the member country associations occupies the chair for a period of two years, on a rotational basis.

The council meets twice per year, normally at the Spring Convention and at the National Honey Show in London, with the remaining meeting by rotation in the member association's country. Council business consists of any matters of common interest to all its members. CONBA provides representation of its membership at the European Union (EU) through two specific committees, COPA and COGECA (COPA – Comite des Organisations Professionelles Agricoles de la CEE); (COGECA Comite de la Cooperation Agricole de la CEE); and the Honey Working Party (HWP).

The Honey Working Party meetings are held at Brussels. This committee liases with the European Commission in relation to apicultural matters concerning the member states of the European Union (EU). These matters are subsequently presented to the European Parliament for its consideration, implementation or revision or rejection. The subsequent approval of such matters results in establishing legislation, government support and possible EC funding relating to the practice of apicultural production in the UK through its membership of the EU.

DARG

DEVON APICULTURAL RESEARCH GROUP

CHAIRMAN, Richard Ball
Stoneyford Farmhouse
Colaton Raleigh
Sidmouth, Devon EX10 0HZ
01 395 567 356

HON SECRETARY, Roger Lacey
Gatchell House
Toadpit Lane
Ottery St Mary EX11 1TR
01404 811 733
devonbees@gmail.com

PUBLICATIONS OFFICER, David Loo
25 Woodlands
Newton-St-Cyres, Exeter
Devon EX5 5BP
0139 285 1472

TREASURER, Bob Ogden
Pennymoor Cottage
Pennymoor, Tiverton
Deven EX16 8LJ
01363 866687

All titles cost £2.50 per copy (post free) from the Publications Officer (tel. 01392 851472). Discounts are available for BBKA affiliated Associations
Please contact the Publications Officer for details

D A R G is an independent group of experienced enthusiastic beekeepers whose primary aim is to collect and analyse data on matters of topical interest which may assist their apicultural education and promote the advancement of beekeeping. At their regular meetings, DARG members discuss various topics in open forum, during which they exchange ideas and information from their personal beekeeping knowledge and experience. They also undertake suitable research projects which further the Group's aims.

TOPICS CURRENTLY BEING UNDERTAKEN
- Use of management (mechanical) methods including shook colonies for varroa control.
- Brood cell size in natural comb.
- A survey of useful bee plants, shrubs and trees in the South West.
- Drone movement between colonies.
In conjunction with Devon BKA
- Survey of Nosema in the County of Devon.
- Survey of drone lying queens in the County of Devon.

PUBLICATIONS AVAILABLE
- **The Beeway Code.** A common sense guide for beginners to help avoid problems with neighbours and produce a safe and peaceful apiary.
- **Seasonal Management**. A useful aid to planning your work effectively
- **Queen Rearing.** Providing detailed help in rearing new queens in order to promote vigorous colonies.
- **The selection of Apiary sites** full of tips for choosing the right sites for your bees.

THE FEDERATION OF IRISH BEEKEEPERS' ASSOCIATIONS

http://www.irishbeekeeping.ie

Secretary: Mr Stuart Hayes,
54 Glenvara Park, Knocklyon, Dublin 16
Tel No 085-1602613, email: fibka.secretary@gmail.com

ANNUAL SUMMER COURSE

The 2014 Beekeeping Summer Course will take place at the Franciscan College, Gormanston, Co Meath from Sunday 27th July to Friday 1st of August. The Guest Speaker will be The guest speaker for 2014 is world renowned Thomas Dyer Seeley, Professor of Biology at the Department of Neurobiology and Behaviour, Cornell University, Ithaca, New York. He is married to Robin and they have two children Saren and Maira

Dr. Seeley graduated from Dartmouth College summa cum laude in 1974 and received his Ph.D. from Harvard University in 1978. He held a postdoctoral fellowship in the Society of Fellows at Harvard until 1980, when he accepted a faculty position at Yale University. He remained there until 1986, when he joined the Department of Neurobiology and Behaviour at Cornell University.

He has pioneered the study of swarm intelligence in social insects through his detailed analyses of how honey bee colonies collectively solve cognitive problems. Through his elegant experimental studies, he has unravelled the complex behavioural mechanisms that enable a honey bee colony to function as a decision-making unit in distributing its foragers among flower patches and in choosing a nesting site. His work has included describing and deciphering several previously mysterious mechano-acoustic signals used by worker bees: the tremble dance, the shaking signal, the worker piping signal, and the stop signal. Seeley's investigations build on the work of Karl von Frisch and Martin Lindauer, but he explores much new ground, asking how a colony as a whole acquires and processes information. In doing so, he discovered fundamental parallels in how a swarm built of bees and a brain built of neurons are organized to make decisions, and thus revealed general principles of organization for building

OFFICERS:
President: Mr Eamon Magee,
222 Lower Kilmacud
Road, Goatstown,
Dublin 14.
Tel No 01-2987611
E-mail eamonma-
gee222@gmail.com

Vice-President:
Mr Gerry Ryan,
Deerpark, Dundrum,
Co Tipperary
Tel No 062-71274/087-
1300751, E-mail ryans-
fancy@gmail.com

P R O: Mr Philip McCabe,
"Sherdara",
Beaulieu Cross,
Drogheda, Co Louth.
Tel No (041-9836159),
E-mail philipmccabe@
eircom.net

Life Vice-Presidents:
Mr P O'Reilly,
11 Our Lady's Place,
Naas, Co Kildare
Tel No (045-897568),
E-mail jackieor@indigo.ie

Mr MI Woulfe,
Railway House,
Midleton, Co Cork
Tel No (021-4631011),
E-mail glenanorehoney@
eircom.net

Mrs Frances Kane,
Firmount, Clane, Co Kildare.
Tel No (087-2450640) or
(045-893150)

Editor: Ms Mary Montaut,
4 Mount Pleasant Villas,
Bray, Co Wicklow.
Tel No 01-2860497. E-mail
yram@connect.ie

Manager: Mr Dermot O'Flaherty,
Rosbeg, Westport,
Co Mayo
Tel No 098-26585/
087-2464045
E-mail:glenderan@anu.ie

Treasurer: Mr Pat Finnegan,
Mullaghmore Road,
Cliffoney, Co Sligo
Tel No 071-9166597/
087-9272692,
email: finnegan@iol.ie

Education Officer:
Mr John Cunningham,
Ballygarron, Kilmeaden,
Co Waterford
Tel No 051-399897/086-
8389108,
email: john3cunningham@
hotmail.com

Summer Course Convenor:
Mr Michael G Gleeson,
Ballinakill, Enfield,
Co Meath.
Tel No 046-9541433/
087-6879584,
email mgglee@eircom.net

intelligent groups. He has written two beautiful books that summarize his discoveries: The Wisdom of the Hive (1995) and Honeybee democracy (2010).

He has travelled and lectured the world over.

In recognition of his scientific work, he has been awarded the Alexander von Humboldt Distinguished U.S. Scientist Prize, a Guggenheim Fellowship, and the Gold Medal Book Award from Apimondia for The Wisdom of the Hive. He has been elected a Fellow of the Animal Behaviour Society and the American Academy of Arts and Sciences. Perhaps his most enduring honour, though, is to have had a species of bee named after him: Neocorynurella seeleyi.

For further information and to secure your place, contact the Summer Course Convenor

Mr Michael G Gleeson, Ballinakill, Enfield, Co Meath.

Tel No 046-9541433/087-6879584, email mgglee@ eircom.net or visit http://www.irishbeekeeping.ie/ gormanston/gormprog2014.html

PUBLICATIONS:

Having Healthy Honeybees - Published by F.I.B.K.A. Editor John McMullan, Ph.D.

The aim of this book is to help beekeepers establish healthy honeybee colonies, assess their condition and take appropriate action. Diseases are dealt with in a concise format to improve readability and are referenced to the latest peer-reviewed research. The book emphasises the importance of proper set-up, involving an integrated approach to health management – in effect a preventative system that comes at little extra cost to the beekeeper

Cost €15 + P & P of €2 each

Bulk buying available to Associations In packs of 10 or 20 books, available at €12 each + P & P of €10 for packs of 10 or 20.

The recommended price is €15 per copy.

It is highly recommended for those doing the various FIBKA Examinations.

Available from Mr Michael G Gleeson, Ballinakill, Enfield, Co Meath.

Tel No (046-9541433) & (087-6879584), E-mail mgglee@ eircom.net

Bees, Hives and Honey - Published by F.I.B.K.A. – Edited by Eddie O'Sullivan

This book has been compiled from writings by some of Ireland's most prominent Beekeepers of the present day. It is an instruction book on beekeeping published as a Millennium project and should prove a modern treatise on the craft of beekeeping and its associated products.

There are over 200 pages, also many photographs and illustrations. Price €12.70 (Paperback) or €19 (Hardback)

Available from Eddie O'Sullivan, Phone: 021-4542614, Email: eosbee@indigo.ie

The Irish Bee Guide – by Reverend J.D. Digges. First published in 1904, it was proclaimed as an excellent book on beekeeping. It also won a place as a notable production in the literary context. It eventually ran to sixteen editions and sold seventy-six thousand copies overall. The name was changed in the second issue to The Practical Bee Guide.

Now, one hundred years later, a decision has been taken to honour this great work. What better way to do it than to re-issue the book as it was in 1904 when it first entered the literary world. The re-print is an exact replica of the original first edition. The price per copy is Hardback 30 and Softback €20

Available from Eddie O'Sullivan, Phone: 021-4542614, Email: eosbee@indigo.ie

An Beachaire – The Irish Beekeeper the monthly organ of FIBKA, subscription €25.00 (Irish Republic), £25 Stg (Northern Ireland/Great Britain) post free from The Manager Mr Dermot O'Flaherty, Rosbeg, Westport, Co Mayo Tel No 098-26585/ 087-2464045

E-mail:glenderan@anu.ie

Readership of the Journal in Northern Ireland carries third party insurance public liability cover up to 6.500, 000 on any one claim and product liability cover up to 6.500, 000 on any one claim, on payment of £5.00 Stg extra.

LIBRARY

The library is owned and controlled by FIBKA. It contains very many valuable books ancient and modern, available to members for return postage only. The Librarian is Jim Ryan, Innisfail, Kickham Street, Thurles, Co Tipperary. Email jimbee1@eircom.net

EDUCATION

The Federation of Irish Beekeepers' Associations (FIBKA) examination system is run by the Education Officer under the direction of the Examination Board; the Board which is made up of members from the FIBKA and the Ulster Beekeepers' Association (UBKA) is appointed by the Executive Council of the FIBKA.

There are seven levels of examination: Preliminary, Intermediate, Senior, Lecturer and Honey Judge Examinations are held during the Summer Course at Gormanston and Preliminary and Intermediate examinations are also held at Provincial Centres.

The Lecturer's examination takes place in the presence of three Examiners, one of whom is the invited Senior Gormanston Summer Course lecturer and also acts as the Extern Examiner.

The Intermediate Proficiency Apiary Practical Examination, the Practical Beemasters Examination and the Apiary Practical component of the Senior Examination are arranged by the Education Officer and take place in the candidate's own apiary during the beekeeping season and are conducted by two Examiners.

The seven levels of examinations for proficiency certificates and their eligibility requirements are as follows:

Preliminary:
For beginners - no prerequisites.
Intermediate:
The Preliminary Certificate of the FIBKA or the BBKA Basic Certificate must be held for at least one year.

Senior:
Intermediate Certificate and at least five years beekeeping experience.
Intermediate Proficiency Apiary Practical
The Intermediate Proficiency Apiary Practical Examination is intended to be part of a stream that will lead to the Practical Beemasters Certificate. The examination is designed to be less "academic" and there are no written examination papers; (it is not part of the Intermediate Certificate Examination).

The examination will take place in the candidate's own apiary and the Examiners will be two Federation Lecturers appointed by the Executive Council. The pass mark is 70%. 20% of the marks scored may be carried forward to the Practical Beemasters Examination

The prerequisites for Intermediate Proficiency Apiary Practical Examination are: the Preliminary Certificate and at least three years' beekeeping experience satisfactory to the Education Board.

The present prerequisites for the Practical Beemasters Certificate are the Preliminary Certificate and at least five years' beekeeping experience satisfactory to the Examination Board - in the future, an additional prerequisite will be the Intermediate Proficiency Apiary Practical Examination.

Practical Beemaster:
Preliminary Certificate and at least five years' beekeeping experience satisfactory to the Examination Board.

Honey Judge:
Intermediate and Practical Beemaster Certificates, successful showing, having obtained a minimum of 200 points at major shows and a record of stewarding under at least four FIBKA Honey Judges.
Lecturer:
Senior Certificate.

Provincial Examinations

Preliminary and Intermediate examinations will be held at provincial centers on the Saturday closest to 6th April (Intermediate) and May 24th (Preliminary). Please note that the minimum number of candidates for a Centre is five for Intermediate and ten for Preliminary. Neighbouring associations may pool their candidates to reach those numbers.

A candidate may sit one Intermediate paper at the Provincial Examination and the other paper at the Summer Course.

The fees for all examinations are valid for the year of application only and are listed on the application forms which may be downloaded from the website. In extreme cases, such as illness (a doctor's certificate must be provided); the examination fee may be held over for one year. There are separate entry forms for the Provincial and Gormanston Summer School Examinations

Fees for Repeat Examinations are the same as for the original examination. Applications to sit the Examinations should be sent to the Education Officer, before the closing dates given above for the Provincial Examinations (applications are however acceptable up to one week after the closing date on payment of a late entry fee which is equal to double the original fee) and before May 1st for the Summer Course Examinations Applications for the Preliminary Examination are also accepted at the Summer Course.

NATIONAL HONEY SHOW

This is held at Gormanston College in conjunction with the annual Beekeeping Course. The Schedule contains 41 Open Classes and 3 Confined classes with €1,000 in prizes. Over 30 Challenge Cups and Trophies are presented for the competition.

Honey Show Secretary: Mr Graham Hall, "Weston", 38 Elton Park, Sandycove, Co Dublin. Tel No (01-2803053) & (087-2406198), E-mail GrahamHall@iolfree.ie

INSURANCE

The limit of indemnity of public liability policy is 6.500, 000 arising from one accident or series of accidents. There is also product liability of €6.500, 000 arising from any one claim. The policy extends to all registered affiliated members whose subscriptions are fully paid up on the 31st December of any one year and whose names are entered in the FIBKA register held by the Treasurer.

FIBKA

ASSOCIATION SECRETARIES

ARMAGH & MONAGH
Mrs. Joanna McGlaughlin
26 Leck Road,
Stewartstown Co Tyrone
BT71 5LS
Tel No 048-87738702/077-
68107984.
joanna.mcg@btinternet.com

Ashford
Mr Michael Giles,
6 The Court, Clonattin
Village, Gorey, Co Wexford.
Tel No 086-8369152.
michaelgiles46@gmail

Ballyhaunis
Mr Gerry O'Neill,
Drimineen South,
Knock Road, Claremorris,
Co Mayo.
Tel No 087 2553533
ballyhaunisbeekeepers@
gmail.com

Banner
Mr Frank Considine,
Clohanmore Cree,
Kilrush, Co Clare,
Tel No 087-6740462,
bannerbees@gmail.com ,

Beaufort
Mr Padruig O'Sullivan,
Beaufort Bar & Restaurant,
Beaufort, Co Kerry.
Tel No 087-258993006,
beaurest@eircom.net

Carbery
Mr Sean O'Donovan,
Drominidy, Drimoleague,
Co Cork.
Tel No 087-7715001.
seanodonovan10@gmail.
com

Co Cavan
Mr Alan Brady,
Shanakiel House,
Drumnagran, Tullyvin,
Co Cavan
Tel No 086-8127920
alan@alanbrady.ie or Info@
alanbradyelectrical.com

Co Cork
Mr Robert McCutcheon,
Clancoolemore, Bandon,
Co Cork.
Tel No 023-8841714.
bob@cocorkbka.org

Co Donegal
Mr Dan Thompson,
Highfield, Loughnagin,
Letterkenny, Co Donegal
Tel No 074-9125894
dthompson@eircom.net

Co Dublin
Mr Liam McGarry,
24 Quinn's Road,
Shankill, Co. Dublin
Tel No 087 2643492.
mcgarryliam@gmail.com

Co Galway
Dr Anna Jeffrey Gibson,
Ballyclery, Kinvara,
Co Galway
secretary@
galwaybeekeepers.com

Co Kerry
Mr Ruary Rudd,
Westgate, Waterville,
Co Kerry.
Tel No 066-9474251.
rrudd@eircom.net

Co Limerick
Mr Gus McCoy,
Mount Catherine Clonlara
Co. Clare
Tel No 087 1390039 :
gusmccoy1@eircom.net

Co Louth
Mr Tom Shaw,
201 Ard Easmuinn, Dundalk,
Co Louth
Tel No 042-9339619/
086-2361286,
tshaw@iol.ie

Co Longford
Mr Joe McEntegart,
Cleanrath, Aughnacliffe,
Co Longford.
Tel No 087-2481340.
josephmcentegart@yahoo.
com

Co Mayo
Ms Helen Thompson,
Graffy, Killasser,
Swinford, Co. Mayo.
Tel No 087-7584835
info@mayobeekeepers.com
or helen.mmooney@gmail.
com

Co Offaly
Mrs Geraldine Byrne,
4 Sheena, Charleville Rd,
Tullamore, Co Offaly
Tel 086-3464545,
loureiro.byrne@gmail.com

Co Waterford
Ms Colette O'Connell,
4 Davis Street, Dungarvan,
Co Waterford
Tel No 058-41910,
coletteoconnell@ymail.com

Co Wexford
Mr John Cloney,
Ballymotey Beg,
Enniscorthy, Co.Wexford.
Tel No 0879801015)
cloney.john@gmail.com

Chorca Dhuibhne
Ms Juli Ni Mhaoileoin,
Burnham, Dingle,
Co Kerry
Tel No 086-8337733,
julimaloneconnolly@gmail.
com

Chonamara
Mr Billy Gilmore,
Maam West, Leenane,
Co. Galway
Tel No 091-571183/087-
7942028,
b.gilmore@
connemarabeekeepers.ie

Digges & Dist
Mr Niall Murphy,
Fenaghbeg, Fenagh,
Ballinamore, Co. Leitrim
Tel 087-7795129/
0719645115
niallmurphy2@hotmail.com

Duhallow
Mr Andrew Bourke,
Pallas, Lombardstown,
Mallow, Co Cork
Tel No 087-2783807.
bourke.andy@gmail.com

Dunamaise
Mr Derek Banim,
Mountain Farm, Killenure,
Mountrath, Co Laois
Tel No 086-0856527,
derekbanim@yahoo.co.uk

Dunmanway
Mr Cormac O'Sullivan,
Forrest Oaks, Forrest,
Coachford, Co Cork.
Tel No 021-7434782
086-4086766
oakforrest@eircom.net

East Cork
Mr C Terry,
Ait na Greine, Coolbay,
Cloyne, Co Cork,
Tel No 021-4652141.
charlesterry@gmail.com

East Waterford
Mr Michael Hughes,
51 Woodlawn Grove, Cork
Road, Waterford
Tel No 051-373461.
waterfordbees@gmail.com

Fingal
Mr John McMullan,
34 Ard na Mara Crescent,
Malahide, Co Dublin
Tel No 01-8450193.
jmcmullan@eircom.net

Foyle
Mr P.J. Costello,
Lr Drumaiveir, Greencastle,
Co Donegal.
Tel No 074-9381303.
pjcost7@eircom.net

Gorey
Ms Cliona Morrish,
Coolkenna, Tullow,
Co. Carlow.
Tel No 086/0874453,
email cliona.morrish@
goreybeekeepers.com

Inishowen
Mr Paddy McDonagh,
Milltown, Carndonagh,
Co Donegal.
Tel No 074-9374881.
paddymcdonough@eircom.
net

Iveragh
Mr Shannon
Ware, 4 Ballinskelligs
Holiday Homes,
Ballinskelligs, County Kerry.
Tel No: 083-3862345 :
research@gamelab.ca

Killorglin
Mr Declan Evans,
Reeks View Lodge, Killorglin,
Co Kerry.
Tel: 087 175 4078, :
declanjevans@gmail.com

Kilternan
Ms Mary Montaut,
4 Mount Pleasant Villas,
Bray, Co Wicklow.
Tel No 01-2860497.
mmontaut@iol.ie

Mid-Kilkenny
Mr Fergal Walsh,
27 The Paddocks, Kells
Road, Co Kilkenny
Tel No 056-7752383 /
086-8402234,
fergalwalsh@eircom.net

New Ross
Mr Seamus Kennedy,
Churchtown,
Feathard-on Sea,
New Ross, Wexford
Tel No 051-397259/
086- 3204236.
seamus.kennedy@yahoo.
co.uk

North Cork
Mr. Eamon Nelligan
3 Carriagroghera
Fermoy Co. Cork.
ciaranneligan@gmail.com

North Kildare
Mr Norman Camier,
34 Lansdowne Park,
Templeogue, Dublin 16.
Tel No 01-4932977/
087-2848938,
norman.camier@gmail.com

Nth Tipperary
Mr Jim Ryan,
"Innisfail", Kickham Street,
Thurles, Co Tipperary.
Tel No 0504-22228.
jimbee1@eircom.net

FIBKA

✉ ☎

Roundwood
Mr John Coleman,
Hillside Cottage,
Roundhill Haven, Clara Beg,
Roundwood, Co. Wicklow
Tel No 087-795 4385,
colemanjkc@gmail.com

Sliabh Luachra
Mr Billy O'Rourke,
Dooneen, Castleisland,
Co Kerry
Tel No 066-7141870,
siobhancorourke@eircom.
net

Sligo/Leitrim
Ms Linda Cheetham,
Shiralees, Montiagh,
Curry, Co Sligo
Tel No 094-9053988.
kenderlyn@googlemail.com
or slba.secretary@gmail.
com

Sneem
Mr Frank Wallace,
Boolananave, Sneem.
County Kerry.
Tel No 086 3522205,
franksneem@hotmail.com

South Donegal
Mr Derek Byrne,
Carrick West, Laghey,
Co Donegal.
Tel No 074-9722340.
dcbyrne@eircom.ie

South Kildare
Mr Liam Nolan,
Newtown, Bagnelstown,
Co Carlow.
Tel No 059-9727281.
liamnolannt@gmail.com

Sth Kilkenny
Mr John Langton,
Coolrainey,
Graiguemanagh,
Co Kilkenny
Tel No 086-1089652,
jjlangton@eircom.net

Sth Tipperary
Mr P J Fegan,
Tickinor, Clonmel,
Co Tipperary.
Tel No 086 1089652,
feganpj@eircom.net

Sth West Cork
Ms Gobnait O'Donovan,
38 McCurtain Hill,
Clonakilty, Co Cork.
Tel No 023-
8833416/083-3069797
gobnaitodonovan@gmail.
com

Sth Wexford
Mr. Dermot O'Grady,
Linden House, Horetown
North, Foulksmills, Co.
Wexford Tel: (051) 565651,
dermaloid@gmail.com

Suck Valley
Ms Anne Towers,
Doonwood,
Mount Bellew, Co Galway.
Tel No 0909-684547/
087-6305714,
annevtravers@gmail.com

The Kingdom
Ms Rebecca Coffey,
75 Ashgrove, Tralee,
Co Kerry
Tel No 066- 7169554,
bexk8@yahoo.co.uk

The Royal Co
Ms Emma Reeves,
Sillogue, Kilberry, Navan,
Co Meath.
Tel No 046-9055808/
087-3590424,
emmafaith.reeves@gmail.
com

The Tribes
Mr Eoghan O'Riordan,
28 Arbutus Avenue,
Renmore, Galway.
Tel No 091-753470/
087-6184132,
landservices@eircom.net

West Cork
Ms Jacqueline Glisson,
Costa Maningi, Derrymihane
East, Castletownbere,
Co Cork. Tel No 086-
3638249,
jglisson@eircom.net

Westport
Mr Dermot O Flaherty,
Rosbeg, Westport,
Co Mayo
Tel No 098 26585/
087-2464045,
info@mayo-westport.com

INTERNATIONAL BEE RESEARCH
ASSOCIATION WEB http://www.ibra.org.uk

IBRA - International Bee Research Association promotes the value of bees by providing information on bee science and beekeeping. This charity was founded in 1949 and is supported by members from around the world. IBRA owns one of the largest international collections of bee books and journals, as well as the Eva Crane / IBRA historical collection and a photographic collection. It operates an online bookshop, publishes its own books and information leaflets, as well as scientific journals.

PUBLICATIONS
Journal of Apicultural Research
A peer reviewed scientific journal that's worldwide and world class. This quarterly publication contains the latest high quality original research from around the world, covering aspects of biology, ecology, natural history and culture of all types of bees.

Bee World
The flagship publication for IBRA members, this quarterly international journal provides a world view on bees and beekeeping. It covers all topics from bee history to the latest findings in bee science.

IBRA BOOKSHOP
The bookshop is accessible via the web site. To support our charitable status IBRA sells a wide range of publications at competitive prices as well as posters, gifts, DVD's and sundries. IBRA is also a publishing house and offers its members a reduction on IBRA products.

CORRESPONDENCE TO:
OPERATIONS DIRECTOR,
Julian Rees
SCIENTIFIC DIRECTOR,
Norman Carreck

Unit 6,
Centre Court,
Main Avenue,
Treforest,
CF37 5YR
029 2037 2409
No fax
www.ibra.org.uk
mail@ibra.org.uk

IBRA

⊠ ☎

MEMBERSHIP
IBRA is proud of its international status and this is reflected by its members who join from all over the world. The membership package now offers more value than ever before: quarterly issues of Bee World, a discount on IBRA publications and online access to a growing back catalogue. For other benefits and the latest information please visit the web site.

Information about all IBRA publications and services can be found via our web site:
www.ibra.org.uk

PLANTS FOR BEES

A Guide to the Plants that Benefit
the Bees of the British Isles

PLANTS
FOR
BEES

A Guide to the Plants that Benefit the Bees
of the British Isles

WDJ KIRK • FN HOWES

W D J Kirk & F N Howes

Foreword by Kate Humble

www.plantsforbees.org

Published and sold by IBRA

IBRA
INTERNATIONAL BEE
RESEARCH ASSOCIATION

Published by IBRA but also available from
Northern Bee Books at
http://www.groovycart.co.uk/beebooks

INIB

✉ ☎

THE INSTITUTE OF NORTHERN IRELAND BEEKEEPERS (INIB)

www.inibeekeepers.com

Annual Conference and Honey Show. 3rd November 2012
Speakers: Keith Delaplane and Ged Marshall
The Village Centre, Hillsborough BT26 6AR

Objectives of the Institute

The Institute is established to advance the service of apiculture and to promote and foster the education of the people of Northern Ireland and surrounding environs without distinction of age, gender, disability, sexual orientation, nationality, ethnic identity, political or religious opinion, by associating the statutory authorities, community and voluntary organisations and the inhabitants in a common effort to advance education, and in particular:

to raise awareness amongst the beneficiaries about bees, bee-keeping and methods of management;

to foster an atmosphere of mutual support among bee-keepers and to encourage the sharing of information and provision of helpful assistance amongst each other.

Affiliation

INIB is affiliated to the British Beekeepers Association.

With 21,100 members the British Beekeepers Association (BBKA) is the leading organisation representing beekeepers within the UK.

As an INIB member, affiliation gives the following benefits.

- BBKA News
- Public Liability Insurance
- Product Liability Insurance
- Bee Disease Insurance available
- Free Information Leaflets to Download
- Members Password Protected Area and Discussion Forum
- Correspondence Courses
- Examination and Assessment Programme
- Telephone Information
- Research Support
- Legal advice
- Representation and lobbying of Government, EU and official bodies.

Events

The Institute holds an annual conference and honey show. The Institute brings to Northern Ireland world renowned expert speakers from USA and Europe to give talks to beekeepers on the latest research and up to date beekeeping methods.

Education

Demonstrations on various topics such as mead making, preparing honey for shows are held during the year.
Courses for honey judges are available.

Honey Bees On Line Studies

INIB has a strong relationship with Professor Jurgen Tautz's of BEEgrouup Biozentrum Universitaet Wuerzburg and his Honey Bee On Line Studies project which continues to develop.

INIB

MEMBERSHIP SECRETARY
Lyndon Wortley
Teemore Grange
224 Marlacoo Rd,
Portadown,
BT62 3TD
Membershipsecretary@
inibeekeepers.com

CHAIRMAN
Michael Young MBE
101 Carnreagh,
Hillsborough
BT26 6LJ
02892689724
chairman@
inibeekeepers.com

Holders of the Institute of Northern Ireland Beekeepers Honey Judge Certificate

001.	MICHAEL BADGER MBE	01132 945879	BUZZ.BUZZ@NTLWORLD.COM
002.	GAIL ORR	02892 638363	GAIL.ORR@BELFASTTRUST.HSCNI.NET
003.	CECIL MCMULLAN	02892 638675	MADELINE.MCMULLAN@HOTMAIL.CO.UK
004.	HUGH MCBRIDE	02825 640872	LORRAINE.MCBRIDE@CARE4FREE.NET
005.	LORRAINE MC BRIDE	02825 640872	LORRAINE.MCBRIDE@CARE4FREE.NET
006.	BILLY DOUGLAS	02897 562926	
007.	MICHAEL YOUNG MBE	02892 689724	CHAIRMAN@ INIBEEKEEPERS.COM
008.	FRANCIS CAPENER	01303 254579	FRANCIS@HONEYSHOW.FREESERVE.CO.UK
009.	MARGARET DAVIES	01202 526077	MARG@JDAVIES.FREESERVE.CO.UK
010.	IAN CRAIG	01505 322684	IAN'AT'IANCRAIG.WANADOO.CO.UK
011.	DINAH SWEET	02920 756483	
013.	LESLIE M WEBSTER	01466 771351	LESWEBSTER@MICROGRAM.CO.UK
014.	REDMOND WILLIAMS	003535242617	EMWILLIAMS@EIRCOM.NET
015.	TERRY ASHLEY	01270 760757	TERRY.ASHLEY@FERA.GSI.GOV.UK
016.	IVOR FLATMAN	01924 257089	IVORFLATMAN@SUPANET.COM
017.	ALAN WOODWARD	01302 868169	JANET.WOODWARD@VIRGIN.NET
018.	DENNIS ATKINSON	01995 602058	DHMATKINSON@TESCO.NET
019	LEO MCGUINNESS	028711 811043	PMCGUINNESS@GLENDERMOTT.COM
020	TOM CANNING		TJCANNING@BTINTERNET.COM
023	ALAN BROWN	01977 776193	ALANHONEYBEES4U@TALKTALK.NET
024	DAVID SHANNON	01302772837	DAVE_ACA@TISCALI.CO.UK

USA
021	ROBERT BREWER		RBREWER@ARCHES.UGA.EDU
022	BOB COLE		
023	ANN HARMAN		

LASI

LASI
LABORATORY OF APICULTURE
AND SOCIAL INSECTS

LABRATORY OF APICULTURE & SOCIAL INSECTS (LASI)

UNIVERSITY OF SUSSEX

FURTHER INFORMATION CONTACT
Francis L. W. Ratnieks,
Professor of Apiculture
Laboratory of Apiculture &
Social Insects (LASI)
Department of Biological &
Environmental Science
University of Sussex, Falmer,
Brighton BN1 9QG, UK

01273 872954 (landline),
07766270434 (mob)
F.Ratnieks@Sussex.ac.uk
www.sussex.ac.uk/lasi

LASI was founded in 1995 and is headed by Francis Ratnieks, who is the UK's only Professor of Apiculture. Prof. Ratnieks received his training in honey bee biology at Cornell University and the University of California in the USA. Whilst in the USA he was also a part-time commercial beekeeper with up to 180 hives used for almond pollination and comb honey production.

From 1995 to 2007, LASI was based at the University of Sheffield. In 2008 Prof. Ratnieks moved to the University of Sussex, which provided LASI with excellent facilities for honey bee research. There is an integrated lab space and offices sufficient for 13 researchers with an adjoining apiary, garden, equipment shed and workshop. There are further apiaries on the university campus and in the surrounding countryside.

LASI is the largest university-based laboratory studying honey bees in the UK and is set up both to undertake research and to train the next generation of honey bee scientists. Undergraduate students receive lectures on honey bee biology and can also do research projects on honey bee biology in their final year and assist LASI research via summer bursaries. Graduate students take a PhD that focuses in a particular area of research. Postdoctoral researchers can learn new skills to complement the training they received during their PhD.

LASI research focuses on both basic and applied questions in bee biology and beekeeping. Basic research areas include communication, foraging, colony organization, nestmate recognition and guarding, and conflict resolution. Applied research areas include improved beekeeping techniques, studies of bee foraging and the value of different plants for honey bees and other pollinators, crop pollination, practical studies of honey bee diseases and their management, and practical measures for bee conservation. Collectively, the LASI team has 80 years of research experience with honey bees.

As well as research and teaching, LASI places great emphasis on outreach and communication. Each year LASI runs workshops, gives talks and writes many outreach articles so that research results are also transferred to beekeepers, gardeners, farmers, land owners, the media, the general public, and policy makers.

Make a date in your diary

April 2014

4 Friday

8:
9:
10:
11:
12:
1:
2:
3:
4:
5:
6:

5 Saturday

BBKA Spring Convention —
Harper Adams University
Lectures, Workshops and Trade Show

6 Sunday

APRIL					
M	T	W	T	F	S	S
	1	2	3	4	5
7	8	9	10	11	12
14	15	16	17	18	19
21	22	23	24	25	2
28	29	30			

THE BRITISH BEEKEEPERS ASSOCIATION · FOUNDED 1874

Spring 2014 Convention

Trade Enquiries – Bob Hunter: bbkasc.trade@virginmedia.com • **General Enquiries** – Tim Lovett: tjl@dermapharm.co.uk

THE NATIONAL DIPLOMA IN BEEKEEPING

The Examinations Board for the National Diploma in Beekeeping was set up in 1954 to meet a need for a beekeeping qualification above the level of the highest certificate awarded by the British, Scottish, Welsh and Ulster Associations.

The Diploma Examination, as designed by the Board, was considered to be an appropriate qualification for a County Beekeeping Lecturer or a specialist appointment requiring a high level of academic and practical ability in beekeeping. It is the highest beekeeping qualification recognised in the British Isles and a high percentage of the past and present holders of the Diploma have given distinguished service to beekeeping education at all levels.

Although the post of County Beekeeping Lecturer has now disappeared, this has merely emphasised the need for some beekeepers to face the challenge of this examination and maintain the high level skills and knowledge needed to keep pace with the increased problems facing all beekeepers at the present time.

The Board consists of representatives from a wide range of organisations and from Government Departments and together form an impressive amalgam of expert knowledge in Beekeeping and Education. Although the National Beekeeping Associations are represented on the Board it is entirely independent of them.

Normally the highest certificate of one of the National Associations is a necessary criterion for eligibility to take the Examination for the Diploma which is held in alternate years. The Written Examination is taken in March, and the Practical, in three sections plus a viva-voce is held in later in the same year.

The Board also organises an annual Advanced Beekeeping Course covering various parts of the syllabus

HON. SECRETARY
Mrs Margaret Thomas
Tig na Bruaich,
Taybridge Terrace,
Aberfeldy, Perthshire,
PH15 2BS.

CHAIRMAN,
Dr David Aston NDB
38 Wressle,
Selby
YO8 6ET
01757 638758

NDB

that are difficult to cover by independent study. Lasting a working week, they cover the main sections of the Syllabus and represent the highest level of training available to British Beekeepers at the present time. The outside lecturers are each acknowledged experts in their particular field. In recent years the Board have been privileged to hold their course at the Fera National Bee Unit at Sand Hutton, York.

In addition the Board organize various short courses at locations in the UK on a number of topics. These are advertised in the bee press and the web site.

For further details regarding the Diploma write, enclosing a stamped A4 SAE to the Secretary, or visit our website: http://www.national-diploma-bees.org.uk/

Those who have gained the National Diploma in Beekeeping

Matthew Allan	Beulah Cullen	*G. Howatson	Bill Reynolds
*Harry Allen	Celia Davis	* Geoff Ingold	Pat Rich
*Harrison Ashforth	Ivor Davis	George Jenner	*Fred Richards
*John Ashton	*Alec S.C. Deans	C. F. Jesson	E. Roberts
Dianne Askquith-Ellis	Clive De Bruyn	Simon Jones	*Arthur Rolt
David Aston	A.P. Draycott	A.C. Kessel	*Jeff Rounce
*John Atkinson	M. Feeley	W.E.Large	Graham Royle
*Miss E.E. Avey	*Barry Fletcher	G.W. Lumsden	J. Ryding
Dan Basterfield	* David Frimston	*Henry Luxton	J.H. Savage
Ken Basterfield	Oonagh Gabriel	A.S. Mcclymont	*Donald Sims
Bridget Beattie	George Gill	J.I. Macgregor	F.G. Smith
*Brig. H.T. Bell	*Reg Gove	Ian Mclean	*George Smith
R.W. Brooke	*Eric Greenwood	Ian A. Maxwell	J.H.F. Smith
Norman Carreck	Pam Gregory	Paul Metcalf	Robert Smith
*Rosina Clark	Anthony R.W. Griffin	J.Mills	*Ken Stevens
Charles Collins	* Robert Hammond	*Bernhard Mobus	*J. Swarbrick
Gerry Collins	Ben Harden	G. N'Tonga	Margaret Thomas
*Tom Collins	C.A. Harwood	*Peter Oldrieve	Adrian Waring
*Robert Couston	*Leslie Hender	Gillian Partridge	Brian Welch
John Cowan	*Alf Hebden	* E.H. Pee	J. Wilbraham
S. J. Cox	*Ted Hooper	I.E. Perera	
Jim Crundwell	Geoff Hopkinson	E.R. Poole	* - deceased

THE NATIONAL HONEY SHOW

www.honeyshow.co.uk
23TH – 25TH OCTOBER 2014.

The Show itself is a wonderful competitive exhibition of all the products of the bee-hive, coupled with an excellent series of lectures, workshops and a wide variety of trade and educational stands.

We recommend that you attend all three days, and suggest that you become a member of the Show – just **£12.00** per annum

For further information, please write to the Hon General Secretary, or Email: showsec@zbee.com or visit our website www.honeyshow.co.uk

HON. GENERAL SECRETARY
REV. H.F CAPENER
1 Baldric Road
Folkestone CT20 2NR

HON TREASURER
C S Mence
27 Acacia Grove
New Malden, Surrey KT3 3BJ

- The National Honey Show is the premier honey show within the United Kingdom.

- Although it is named the "National Honey Show", it includes a strong international element.

- As well as the competitive content of the Show, there is also a full programme of lectures and workshops.

- In the Sales Hall, all the major traders and educational organisations are present.

- Further information is readily available on the website www.honeyshow.co.uk or from the Hon General Secretary showsec@zbee.com

NIHBS

✉ ☎

The Native Irish
Honey Bee Society
Apis mellifera mellifera

NATIVE IRISH
HONEY BEE SOCIETY

CHAIRPERSON:
MR. PAT DEASY
chairperson@nihbs.org

SECRETARY:
MRS. COLETTE O CONNELL
secretary@nihbs.org

TREASURER:
MR. SEÁN Ó FEANNACHTA
treasurer@nihbs.org

PUBLIC RELATIONS OFFICER:
MS. AOIFE NIC GIOLLA CODA
pro@nihbs.org

WEBMASTER:
JONATHAN GETTY
webmaster@nihbs.org
www.nihbs.org

FACEBOOK PAGE
www.facebook.com/
native-irish-honey-bee-society

What is the Native Irish Honey Bee Society?
NIHBS was established in November 2012 by a group of beekeepers who wish to support the various strains of Native Irish Honey Bee (Apis mellifera mellifera) throughout the country. It is a cross border organisation and is open to all. It consists of members and representatives from all corners of the island of Ireland.

Aims and Objectives -
To promote the conservation, study, improvement and re-introduction of Apis mellifera mellifera (Native Irish Honey Bee), throughout the island of Ireland.
- To establish areas of conservation throughout the island for the conservation of the Native Honey Bee.
- To promote formation of bee improvement groups.
- To provide education on bee improvement and to increase public awareness of the native honey bee.
- To act in an advisory capacity to groups and individuals who wish to promote it.
- To co-operate with other beekeeping organisations with similar aims.
- To seek the help of the scientific community and other stake holders in achieving our aims and objectives.

164

NIHBS - Plans for the future -
Liaise with groups interested in Native Honey Bees
• Will apply for funding -
• Will help to co-ordinate projects -
• Will raise awareness to beekeepers and the public about Native Honey Bees -
• Talks and lectures

Why Join NIHBS?
• Information on beekeeping events around Ireland – North and South
• queen rearing workshops, talks and lectures.
• Information on how to obtain Native Honey Bees
• Conference discounts
• Discounted entrance fees to events run by NIHBS
• Eligibility to schemes coordinated by NIHBS
• A network of beekeepers interested in our native honeybee

1 years membership costs 20 euro or 20 pounds sterling.
A membership form can be downloaded from website and sent to treasurer or payment can be made on website via paypal.

ROTHAMSTED RESEARCH

www.rothamsted.ac.uk

ROTHAMSTED RESEARCH
Department of AgroEcology,
Rothamsted Research
Harpenden,
Hertfordshire.
AL5 2JQ

STAFF
Dr Alison Haughton
Dr Beth Nicholls
Dr Stephan Wolf
Dr Jason Lim
Dr Samantha Cook
Jenny Swain
Jonathan Carruthers
(PhD student)
Steve Kennedy
Sue Bird

The Rothamsted site provides a unique working environment with specialist modern equipment facilitating research on plant and microbial metabolites, molecular biology and synthetic and analytical chemistry. There is an experimental farm for complex field experiments, and there is a suite of glasshouses, controlled environment facilities, an insectary and a state-of-the-art bioimaging suite housing three new electron microscopes and a confocal laser scanning microscope. Experimental design and analysis are backed up by excellent statistical, computing and library support.

BEE BEHAVIOUR AND POLLINATION ECOLOGY

We are investigating the interaction between bees, crops and the agricultural environment. The spatial and temporal foraging behaviour of honey bees and bumble bees within agricultural areas is being compared. Harmonic radar is being used to track flying bees, and other pollinators such as butterflies, to obtain new information about their flight paths, forage ranges, food preferences and orientation mechanisms.

An integrated model for predicting bumblebee population success and pollination services in agro-ecosystems will be developed by Rothamsted and colleagues at the Environment & Sustainability Institute at the University of Exeter and the University of Sussex, and will provide a powerful tool for shaping recommendations for land managers and policy makers for the sustainable spatial management of pollination within arable and horticultural production systems.

Various qualities of different varieties of crops (oilseed rape and short rotation coppice willows) as important resources for bees are being investigated. The nutritional value of the nectars and pollens, effects on bee fitness and behaviour are key areas of interest.

HONEY BEE PATHOLOGY

Rothamsted's research on the natural history and epidemiology of the infections and parasites of bees has had wide international recognition. However, research on honey bee pathology is currently suspended due to changes in funding available from Defra for bee health. Over the last 20 years, this work focused on *Varroa destructor* and the losses caused by honey bee virus infections that the mite transmits. In a collaborative project with Horticulture Research International (at University of Warwick), investigating potential biological control agents of *V. destructor*, the research identified and characterised fungal pathogens which are active against the mite but which are relatively safe for bees and other beneficial insects. Biological control offers an environmentally acceptable approach to the problem that could have considerable economic benefits, and we are actively seeking funding to continue this work.

An Insect Pollinator Initiative funded project is assessing the impact of emergent diseases, including the *Varroa* associated Deformed wing virus, and the Microsporidian *Nosema ceranae* on the flight performance and orientation ability of honeybees and bumblebees and its consequences for bee populations.

HARMONIC RADAR

The use of harmonic radar in insect behaviour studies has been pioneered at Rothamsted. A transponder weighing just a few milligrams fitted to the thorax of bees picks up the interrogation radar signal and immediately emits a signal at a different frequency, which is then received by the radar. A recently awarded European Research Council grant will now enable cutting-edge development of the harmonic radar to allow us to collect data for entire adult life-spans and foraging ranges for multiple individuals of bee species, thus allowing us whole new insights into bee behaviour and pollination ecology.

INFORMATION EXCHANGE

Expertise in bee research is drawn upon by scientific colleagues world-wide and there are research links with institutes and universities in this country and abroad. Research findings are published in scientific journals but popular articles are also written for the beekeeping and agricultural press. Effective communication of our science by staff members is delivered via a vigorous programme of lectures presenting to national and local beekeeping associations and participation in various public media, including BBC Horizon.

FUNDING

Rothamsted receives funds for research from the Biotechnology and Biological Sciences Research Council, through competitions and contracts from the Department for Environment, Food and Rural Affairs, the European Community, from Levy boards, commercial and other organisations. The support of the bee research programme in recent years by grants from the British Beekeepers Association, C. B. Dennis British Beekeepers Research Trust, the Eastern Association of Beekeepers and the Bedfordshire, Cambridgeshire, Norfolk, St Albans and Hertfordshire and High Wycombe Beekeepers Associations is gratefully acknowledged.

For more information visit: **http://www.rothamsted.ac.uk**

THE SCOTTISH BEEKEEPERS' ASSOCIATION

AIMS OF THE ASSOCIATION
- publish a monthly magazine
- maintain the Moir Library in Edinburgh
- conduct examinations in the art of beekeeping
- provide insurance and a compensation scheme for members

EDUCATION
The SBA arranges courses and awards certificates to successful candidates in the Scottish Basic Beemaster, Expert Beemaster, Honey Judge and Microscopy Examinations. It also actively promotes beekeeping by informing the public, especially the young, about bees and their benefits to the environment.

INSURANCE AND THE COMPENSATION SCHEME
All members of the SBA have insurance against Public Liability. The SBA Compensation Scheme is restricted to bee colonies located in Scotland and allocates part-replacement value for damage by vandalism, fire, theft and certain brood diseases.

LIBRARY
The SBA Moir Library in Edinburgh has one of the world's finest collection of beekeeping books. A library card is issued annually to every member who can borrow books at the cost of return postage only. Details may be obtained from the Library Convener.

MARKETS
Advice is given on all aspects of marketing honey products at appropriate times. Suggested bulk, wholesale and retail prices are notified in the magazine.
PUBLICATIONS

GENERAL SECRETARY
Mrs. Bronwen Wright
20 Lennox Road
Edinburgh EH5 3JW
0131 552 3439
secretary@
scottishbeekeepers.org.uk

HON PRESIDENT
The Rt. Hon. Earl of Mansfield D.L, J.P
Scone Palace
Perth PH2 6BE

HON. VICE PRES,
Iain F Steven
4 Craigie View
Perth
PH2 0DP
01738 621100
Ian Craig
30 Burnside Ave, Brookfield,
Johnstone, Renfrewshire,
PA5 8UT
01505 322684
beekeeper30@btinternet.
com

HON. LIBRARIAN
Mrs. Margaret M. Sharp
City Librarian, City Library
George IV Bridge, Edinburgh

HON. LEGAL ADVISER,
Taggert, Meil & Mathers
20 Bon Accord Sq,
Aberdeen
01224 588020

SBA

✉ ☎

INDEPENDENT EXAMINER
Lynne Ramsay
Lynne Ramsay,
11, Chancelot Terrace,
Edinburgh EH6 4SS
0131 552 8218

- The Scottish Beekeeper is published monthly and sent post free as part of the annual membership fee of £30 payable to the Membership Convener.
- Introduction to Bees and Beekeeping is £6.00 plus postage and may be obtained from the Advertising and Publicity Convener.

PUBLICITY

Members can purchase the Association tie, lapel badge, car sticker etc. Details may be obtained from the Advertising and Publicity Convener.

SHOWS

Two major annual honey shows are held in Scotland.

A honey competition and show with educational displays is held at the Royal Highland Show, Ingleston, Edinburgh in June and the Scottish National Honey Show is conducted at the Dundee Food and Flower Festival in September. Other Honey Shows are run in Ayr, Fife, Inverness, Turiff and at many other locations in Scotland as organised by Local Associations.

Executive Committee

PRESIDENT,
Phil McAnespie
12 Monument Road
Ayr KA7 2RL
01292 885660
membership@
scottishbeekeepers.org.uk

VICE PRESIDENT,
Mrs. Bronwen Wright
20 Lennox Road, Edinburgh
EH5 3JW
0131 552 3439
secretary@
scottishbeekeepers.org.uk

IMM. PAST PRES,
Ian Craig
30 Burnside Avenue
Brookfield, Johnstone
Renfrewshire PA5 8UT
01505 322684
beekeeper30@
btinternet.com

GENERAL SEC
Mrs. Bronwen Wright
20 Lennox Road, Edinburgh
EH5 3JW
0131 552 3439
secretary@
scottishbeekeepers.org.uk

SBA CO-ORDINATOR,

TREASURER, IAN ROLLO,
36 Newton, Cupar, Fife
KY15 4DD
01334 650 836
ianrollo@yahoo.co.uk

EDITOR, SCOTTISH BEEKEEPER,
Nigel Southworth
47 Middleton Road, Uphall,
Edinburgh, EH52 5DF
01506 865762
editorscottishbeekeeper@
gmail.com

CONVENERS OF STANDING COMMITTEES

MEMBERSHIP CONVENER
P. McAnespie
12 Monument Rd.Ayr
KA7 2RL   01292 885660
membership@
scottishbeekeepers.org.uk
INSURANCE & COMPENSATION
C. Irwin
55 Lindsaybeg Rd
Chryston, Glasgow
G69 9DW
0141 7791333
ceirwin@talktalk.net
ADVERTISING & PUBLICITY
Miss E Brown
Milton House, Main Street
Scotlandwell, Kinross
KY13 9JA   01592 840582
honeybees@onetel.com
EDUCATION, Alan Riach
 Woodgate, 7 Newlands Ave,
Bathgate
EH48 1EE
01506 653839
alan.riach@which.net
PROMOTION OF BEEKEEPING
CONVENER

SHOWS, Miss E Brown
Milton House, Main Street
Scotlandwell, Kinross
KY13 9JA
01592 840582
honeybees@onetel.com
LIBRARY, Mrs Una Robertson
13 Wardie Ave
Edinburgh
EH5 2AB
una.robertson@btinternet.
com
MARKETS, Margaret Thomas
Tig na Bruaich, Taybridge
Terrace, Aberfeldy,
Perthshire PH15 2BS
01887 829 710
zyzythomas@waitrose.com
BEE HEALTH, Phil Moss
Ealachan Bhana
Clachan Seil
Oban
PA34 4TL
01852 300383
phil.moss@dsl.pipex.com
ICT CONVENER,
Alasdair Joyce
Manachie Lodge.
Dallas Dhu
Forres
IV36 0RR
01309 671288
webmaster@
scottishbeekeepers.org.uk

AREA REPRESENTATIVES
NORTH,
Mrs Sheila Barnard
Viewmount, Tobermory,
Isle of Mull
PA75 6PG
01688 302008
tim-barnard@lineone.net
EAST, JOHN COYLE
Rose Cottage, Burnton,
By Kippen, Stirling
FK8 3JL
07774 266 540
 info@beekeepinginscotland.
co.uk
WEST, Mike Thornley
Glenarn House, Glenarn
Road, Rhu, Helensburgh
G84 8LL
01436 820493
masthome@dsl.pipex.com
ABERDEEN AND MORAYSHIRE,
Dr Stephen Palmer
Fintry School House,
Fintry, near Turiff
AB53 5RN
01888 551367
palmers@fintry.plus.com

SBA

✉ ☎

S.B.A LECTURERS *Addresses in SBA Honey Judges List

All those listed may claim expenses except G. Sharpe, Scottish Bee Inspectors and Stephen Sunderland, all funded by SGRPID. All speakers accompany talks with visual aids

* **MISS. E. BROWN** (General)
01592 840542

* **M BADGER** (General)
0113 2945879

* **I. CRAIG** (General)
01505 322684

A.B. FERGUSON
(General, Varroa)
Firparkneuk. Kirtlebridge
Lockerbie   DG11 3LZ
01461 500322

* **C. IRWIN** (General)
0141 7791333

* **DR. F. ISLES** (Bee diseases)
01382 370 315

M.M. PETERSON
(Bee genetics)
Balhaldie House,
High street, Dunblane
FK15 0ER
01786 822093

G. SHARPE (SAC) (Varroa
Management: My apiary
management system)
Apiculture Specialist
Life Science Technology
Group, SAC Auchincruive
Ayr KA6 5HW
01292 525375

Mrs M Thomas (General)
Tighnabraich, Taybridge
Terrace, Aberfeldy
Perthshire PH15 2BS
01887 829710

DR G RAMSAY (Beekeeping
on the Internet / Can Bees
fight Varroa?)
Parkview, Station Road
Errol, Perth   PH2 7SN
01821 642385

A RIACH
(Beehives through the
Ages)
Woodgate, 7 Newland Ave
Bathgate
EH48 1EE
01506 653839

PAUL GIBSON,
7 Shielswood Court,
Galashiels, Selkirkshire
TD1 3RH
01896 750110
paulalisongibson@
btinternet.com

BRYCE REYNARD, 39 Old Mill
Lane, Inverness
IV2 3XP
01463 225887
elizabethreynard@
btinternet.com

SCOTTISH BEE INSPECTORS,
SGRPID, P Spur, Saughton
House, Broomhouse,
Edinburgh
EH11 3XD
0300 244 6672
beesmailbox@scotland.gsi.
gov.uk

DR PETER STROMBERG,
21 Woodside, Houston,
Renfrewshire,
PA6 7DD
01505 613 830
pstromberg1@aol.com

STEPHEN SUNDERLAND,
Lead Bee Inspector,
SGRPID, P Spur,Saughton
House, Broomhouse,
Edinburgh
EH11 3XD
0300 244 6672
steve.sunderland@
scotland.gsi.gov.uk
beesmailbox@scotland.gsi.
gov.uk

DR DAVID WRIGHT,
20 Lennox Row,
Edinburgh
EH5 3JW
0131 552 3439
bdwright20lr@btinternet.
com

SBA

✉ ☎

MEMBER ASSOCIATIONS AND THEIR SECRETARIES

ABERDEEN, Rosie Crighton
29 Marcus Cresc
Blackburn, Aberdeen
AB21 0SZ
01224 791181
rosie@crighton- findlater.
fsbusiness.co.uk
**ARRAN BEE GROUP, W K
McNeish** Seafield, Kildonan,
Isle of Arran
KA27 8SE
01770 820357
wmcnsh@aol.com
AYR, Mrs L Baillie
Windyhill Cottage
Uplands Rd, Sundrum
Ayre, KA6 5JU
01292 570659
lbaillie@sundrum.demon.
co.uk
BORDER, Liz Howell
Oatlands, Houndridge,
Kelso
TD5 7QN
01573 470747
kevhwl@aol.com
BUTE, Alison Cross
Marionslea, Minister's Bray,
Rothsay, Isle of Bute
PA20 9BG
01700 504627
alison.cross2@virgin.net
**CADDONFOOT,
James & Julia Edey**
West Water, Bedrule,
Hawick, Roxburghshire,
TD9 8TD
01450 870400
jamesedey@googlemail.
com

**COVINGTON AND THANKERTON,
Angus Milner-Brown**
Covington House,
Covington Road, Biggar
ML12 6NE
01899 308024
angus@therathouse.com
**CLYDE AREA, Mr George
Morrison**
102 Woodside Ave
Bearsden G61 2NZ
0141 942 9419
COWAL, Brian Madden
123a Alexandra Parade
Dunoon, PA23 8AW
01369 703317
brian_maden@btinternet.
com
DINGWALL, Alpin Stewart
Rowan Cottage,
Fasaig, Torridon by
Achnasheen, Ross-shire
IV22 2EZ
01445 791450
dingwall.beekeeping@
googlemail.com
**DUNBLANE & STIRLING,
Fiona Fernie**
Greystones Dunira,
By Comtie
PH6 2JZ
01764 679152
secretary@
dunblanebeekeepers.com

DUNFERMLINE & WEST FIFE
Dr T Scott
Grange Farmhouse
Grange Rd, Dunfermline
KY11 3DG
01383 733125
somatocs@gmail.com
**EAST OF SCOTLAND,
Edward Summerton**
Mains of Airlie Farmhouse,
Kirriemuir, Angus
DD8 5NG
secretary@
eastofscotlandbeekeepers.
org.uk
**EAST LOTHIAN, Deborah
Mackay**
5 Goshen Farm Steading,
Musselburgh, East Lothian
EH21 8JL
0131 665 8939
eastlothianbeekeepers@
googemail.com
EASTER ROSS, Colin Ridley
Stirling Cottage,
Lamington Park, Kildray,
Ross-shire
IV18 0PE
01862 842410
colinr031@googlemail.com
EASTWOOD, Jackie Reid
6 Sutherland, Brancumhall,
East Kilbride,
Glasgow
G74 3DL
01355 279901
eastwoodbeekeepers@
hotmail.co.uk

SBA

✉ ☎

EDINBURGH & MIDLOTHIAN
P Steven
Eastercowden Cottage
Dalkeith, Midlothian
EH22 2NS
07703 528801
porrsteven@yahoo.co.uk
FIFE, Janice Furness
The Dirdale, Boarhills
St. Andrews, Fife KY16 8PP
01334 880 469
jcfurness@dirdale.fsnet.
co.uk
FORTINGALL, Mrs. Jo Pendleton,
Lilac Cottage
Old Bridge of Tilt by
Pitlochry
PH18 5TP
01796 481 362
d.h.pendleton@btinternet.
com
GLASGOW DISTRICT, Mhairi Neill
3 machan Ave, Larkhall,
ML9 2HE
01698 881602
glasgowbeeksec@hotmail.
co.uk
HELENSBURGH, Gordon Smith
The Moorings, Ferry Road,
Rhu G84
07980 578206
secretary@helensburghbees.
com
HONEYPOTZ (WEST LOTHIAN),
Dave Gillan
34 Gardner Crescent,
Whitburn,
EH47 0PE
01501 744817
honeypotzbeekeeping@
live.co.uk
INVERNESS-SHIRE
Julia Moran
ibasecretary@hotmail.co.uk

KELVIN VALLEY, I Ferguson
4 South Glassford Street
Milngavie G62 6AT
0141 956 3963
jeanian@ferguson2007.plus.
com
KILBARCHAN AND DISTRICT
I. Craig
30 Burnside Ave
Brookfield
Johnstone PA5 8UT
01505 322684
beekeeper30@btinternet.
com
KINTYRE & MID ARGYLL,
Mike Stanesby
c/o The Pheasntry,
Balinakill, Clachan,
Tarbert, Argyll
PA29 6XL
01880 740647
mike.stanesby@clachan.
argyll-bute.sch.uk
KILMARNOCK & DISTRICT
J. Campbell
North Kilbryde House
Stewarton
Kilmarnock KA3 3EP
01560 482489
john.d.campbell@talktalk.net
KIRRIEMUIR,
'Disbanded'
LARGS & DIST, Evelyn Mackie
Skelmorlie Mains Cottage,
Skelmorlie, Ayrshire
PA15 5EU
07952 720247
e.j.mackie@hotmail.co.uk
LOCHABER, Rev Kate Atchley
Anasmara, Mingarry
Acharacle, Inverness-shire
PH36 4JX
01967 431420
contact@kateatchley.co.uk

MORAY, YVONNE STUART
The Cottage, Norht
Darkland, Lhanbryde, Moray
IV30 8LB
01343 842317
secretary@
moraybeekeepers.co.ukcom
MULL, Mrs. S. Barnard
Viewmount, Tobermory
Isle of Mull PA75 6PG
01688 302008
tim-barnard@lineone.net
NAIRN & DISTRICT, John Burns
Woodlands, Cawdor Road
Nairn IV12 5EF
01667 454887
jandjburns@hotmail.com
NEWBATTLE
(FORMERLY LAMANCHA)
Joyce Jack,
23 South Park West, Peebles
EH45 9EF
01721 722444
joycecjack@aol.com
OBAN & DISTRICT,
Phil Moss
An Isean Eala
Clachan Seil
Oban PA34 4TL
01852 300383
aniseaneala@btinternet.com
OLRIG and District, Robin Inglis
Roadside Skirza
Freswick, Wick KW1 4XX
01955 611260
gailinglis@btinternet.com
ORKNEY, Sue Spence
Alton House, Berstane Road,
Kikwall, Orkney
KW15 1NA
01856 873920
bs3920@yahoo.com

PEEBLES-SHIRE,
Amanda Clydesdale
20 Kingsmeadows Gardens
Peebles EH45 9LB
01721 720563
amanda.clydesdale@
btinternet.com
PERTH AND DISTRICT ,
Linda Legget
2 Acharn, Perth
PH1 2SR
01738 580024
info@
perthanddistrictbeekeepers.
co.uk
SKYE & LOCHALSH,
Rod Haswell
Tigh na Chroisean,
4 Black Park,
Boradford, Isle of Skye
IV49 9DE
01471 822338
rod@thehaswells.com
S. OF SCOTLAND, Edith
Reyntiens 40 George Street,
Dumfries DG1 1EH
01387 266583
fergiearchie@tiscali.co.uk
fergiearchie@tiscali.co.uk

SUTHERLAND, Sue Steven
Mulberry Croft, 2 East
Newport,
Berriedale Caithness KW7
6HA
01539 751 245
mulberrycroft607@
btinternet.com
WEST'N GALLOWAY,
Linda Robertson
Craigenveoch Farm,
Glenluce, Newton Stewart,
DG8 0LD
07825 51 4 726
lindaglenmheran@aol.com
WEST LINTON & DISTRICT
D. Stokes
100 Main Street, Roslin
Midlothian EH25 9LT
0131 440 3477
wlbka@live.co.uk
WESTERN ISLES,
Martin Johnstone
3 Upper Bayble Point,
Isle of Lewis, Western Isles
HS2 0QH
martinjohnstone1@tiscali.
co.uk

Freuchie BKA
disbanded

175

SBA

✉ ☎

SBA ACTIVE HONEY JUDGES

M BADGER
Kara, 14 Thorn Lane,
Roundhay,
Leeds
LS8 1NN

MISS E. BROWN
Milton House, Main Street,
Scotlandwell
Kinross
KY13 9JA
01592 840582

P.J. BROWNE
The Rowan Tree, Gairlochy
Spean Bridge
Inverness-shire
PH34 4EQ
01397 712730

M. CANHAM
Whinhill Farm House
by Cawdor, Nairn IV12 5RF
01667 404314

I. CRAIG
30 Burnside Avenue
Brookfield, Johnstone
Renfrewshire,
PA5 8UT
01505 322684

H DONOHOE
7 Grant Road
Banchory
AB31 5UW
01330 823502

C. E. IRWIN
55 Lindsaybeg Road
Chryston, Glasgow
G69 9DW
0141 7791333

DR F. ISLES
"Gardenhurst",
Newbigging Broughty Ferry
Dundee
DD5 3RH
01382 370315

P MATHEWS
MRS C MATHEWS
4 Annanhill
Annan, Dumfries-shire
DG12 6TN
01461 205525

MS B L MCLEAN
Upper Flat, 2 Invererne Rd,
Forres
IV36 1DZ
01309 676316

C. WEIGHTMAN
Shilford, Stocksfield,
Northumberland
NE43 4HW
01661 842082

C. WILSON
Cedarhill, Auchencloch,
Banknock, Bonnybridge
FK4 1VA
01324 840227

DR D WRIGHT
MRS B WRIGHT
20 Lennox Row
Edinburgh
EH3 5JW
0131 552 3439

M. YOUNG
101 Carnreagh,
Hillsborough
County Down
N. Ireland
BT26 6LJ
0289 268972

ULSTER BEEKEEPERS' ASSOCIATION

www.ubka.org

OBJECTS OF THE ASSOCIATION

Tho objocts of the Association are to unite beekeepers for their mutual benefit to serve the best interests of beekeeping by all means within its power and to foster its healthy development.

For the purpose of achieving these objects the Association will:

- promote the formation of local Beekeepers' Associations
- disseminate information and advice about beekeeping
- provide examination facilities in the craft of beekeeping
- encourage maintenance and improvement of the beekeeping environment.

EDUCATION

In conjunction with the College of Agriculture, Food & Enterprise (CAFRE), the U.B.K.A. assists in organising classes for Preliminary, Intermediate and Senior Certificate Examinations in Beekeeping following the syllabus of the Federation of Irish Beekeepers' Associations (FIBKA).

INSURANCE

Affiliated local Associations and their individual members have access to the UBKA group public and product liability insurance scheme.

APIARY SITES

Almost all twelve local Associations and CAFRE's Greenmount Campus have access to apiary sites and, for some sites, access to observation houses provided with help from Leader 2 funding, for use in demonstrating and promoting good practice to members, schools and other interested groups.

PRESIDENT,
David Wright
24 Quarry Road
Lisbane, Comber,
Newtownards
Co Down BT23 5NF

SECRETARY,
Brian Richardson
Agho,
305 Lattone Road,
Belcoo,
Fermanagh
BT93 5ES

TREASURER,
Gail Orr.
64 Ballycrune Rd.,
Hillsborough,
BT26 6NH

LECTURERS
Vanessa Drew,
40 Lacken Rd.,
Ballyroney,
Banbridge,
Down BT32 5JA

Jim Fletcher
26 Coach Road, Comber
Co.Down. BT23 5QX

UBKA

✉ ☎

Lecturers continued
Ethel Irvine
2 Laragh Lee
Ballycassidy
ENNISKILLEN
BT94 2JT

Lorraine McBride
11, Ballyloughan Park
Ballymena, Co.Antrim,
BT43 5HW

Rev Sam Millar
41 Rectory Park
Garvagh, COLERAINE
Co Londonderry
BT51 5AJ

Norman Walsh
43, Edentrillick Rd
Hillsborough, Co. Down
BT26 6PG

HONEY SHOWS

Local Associations stage honey shows throughout Northern Ireland. The Northern Ireland Honey Show, hosted by the Belfast City Parks Department, is held annually in September in the Botanic Gardens Belfast.

CONFERENCE

The 70th UBKA Annual Conference will be held on 7 – 8th March 2014 at CAFRE's Greenmount Campus, Antrim. Contact the U.B.K.A. Conference Secretary at 07871 161303 and www.ubka.org for details.

SECRETARIES OF ASSOCIATIONS

Belfast,
Alan Rea
12 Kirkliston Drive
Belfast
BT5 5NX

Belfast, Jonathan Getty
80 Locksley Park,
Belfast,
BT10 0AS

Clogher Valley, Chester
Roulston
10 Ednagee Rd,
Garvetagh, Castlederg,
Co. Tyrone
BT81 7QF

Derry City, Jen Simpson
4 Cullinean Manor,
Redcastle,
Nr Lifford
Co. Donegal

Dromore, Patrick Lundy
116 Dromore Road,
Ballynahinch,
Co.Down
BT24 8HK

East Antrim, Stephen Robinson
53 Wellington Ave.,
Larne, Co Antrim
BT40 1EH

Fermanagh, Joanne McNulty
30a Cornahiltie Rd,
Gortnalee, Belleck,
Co. Fermanagh
BT93 3 AU

Killinchy, Dawn Stocking
Ballycruttle House,
7 Tullynaskeagh Road
Downpatrick
Co. Down
BT30 7EJ

Mid Antrim, Angela Morrow
23 Beechwood Drive
Ahogill
Ballymena
BT42 1NB

Mid Ulster, Anne Milligan
61 Blackisland Road ,
Annaghmore,
Portadown
BT62 1NE

Randalstown, Susie Hill
7 Nutts Corner Road,
Crumlin,
Antrim
BT29 4BW

Roe Valley, Sandra Logan
22 Knocknougher Road,
Macosquin,
Coleraine,
Co. Londonderry.

Rostrevor & Warrenpoint,
Christina Joyce
"The Grange",
1 Mourne Park, Kilkeel,
Co Down
BT34 4LB

Honey Judges
Jim Fletcher
26 Coach Road,
Comber
BT23 5QX

Michael Young
Mileaway, Carnreagh Road
Hillsborough
Co. Down
BT26 6LJ

Norman Walsh
43 Edentrillick Rd
Hillsborough
Co. Down
BT26 6NH

CYMDEITHAS GWENYNWYR

CYMRU WELSH BEEKEEPERS' ASSOCIATION

AMCANION Y GYMDEITHAS / AIMS OF THE ASSOCIATION
- Promote and develop beekeeping in Wales
- Conduct examinations in beekeeping
- Liaise with organisations and bodies for the benefit of beekeeping in Wales

AELODAETH UNIGOL / INDIVIDUAL MEMBERSHIP
Individual membership of the WBKA is provided for persons who do not live within the areas of branch associations, and wish to support the association. Information relating to benefits and facilities provided for individual members is available from the Individual Membership Secretary.

ARHOLIADAU / EXAMINATIONS
The Examinations Board conducts six grades of examinations: Junior, Primary, Intermediate, Practical, Honey Show Judges, Senior. Information is available from the Examination Board Secretary.

Candidates following the Duke of Edinburgh Award Scheme may receive information regarding the inclusion of beekeeping as a course submission from the Examinations Secretary.

CYNHADLEDD/ CONVENTION
At the Royal Welsh Agricultural Society's Showground, Llanelwedd. This event is normally held during Late March/ Early April. Information relating to this event is available from the convention secretary.

YSWIRIANT / INSURANCE
All individual and fully paid up members of beekeeping associations affiliated to WBKA are covered against 'Public and Product' liability claims. All affiliated associations are covered against public liability during conventions officially organised by the association.

YSGRIFENNYDD / SECRETARY
John Page
The Old Tannery
Pontsian
Llandysul
Ceredigion
SA44 4UD
secretary@WBKA.com

LLYWYDD/PRESIDENT
David Culshaw,
9 Ash Grove, Llay,
Wrexham LL12 0UF
president@wbka.com

CADEIRYDD/CHAIR
Jenny Shaw
Llwyn Ysgaw,
Dwyran, Llanfairpwll,
Anglesey LL61 6RH
chair@wbka.com

IS-GADAIRYDD/
VICE CHAIR
Position Vacant
depchair@wbka.com

WBKA/CGC

✉ ☎

TRYSORYDD/TREASURER
Vincent Frostick

The WBKA Individual Membership benefits include cover under the BDI Scheme against the loss, due to foul brood diseases, of a minimum number of stocks (determined by BDI). Affiliated Associations provide this cover for their members.

GWEFEISTR/WEBMASTER
AND GOLYGYDD/EDITOR
Sue Closs
editor@wbka.com

LLYFRGELL / LIBRARY

The reference sections of all county libraries in Wales have details of the names and addresses of Secretaries of Associations affiliated to WBKA.

IS-OLYGYDD (ERTHYGLAU
CYMRAEG)/SUB EDITOR
Dewi Morris Jones
Llwynderw, Bronant
Aberystwyth SY23 4TG
(01974 251264)

Books on beekeeping can be borrowed from county, branch and mobile libraries. The Library, Ffordd y Bala, Dolgellau LL40 2YS, has been nominated to stock beekeeping books.

Members of associations affiliated to IBRA may borrow books/documents from its library.

GWENYNWYR CYMRU - The Welsh Beekeeper

ARHOLIADAU/EXAMINATIONS
Dinah Sweet
Graig Fawr Lodge
Caerphilly
CF83 1NF
education@wbka.com

A publication of the Welsh Beekeepers Association, giving news and views of beekeeping and related subjects. Articles and advertisements enquiries should be sent to the Editor. Articles written in Welsh should be sent to the Sub Editor. Gwenynwyr Cymru is provided free to members of Affiliated Associations and Individual Members. Information regarding subscriptions is available from the Individual Membership / Subscription Secretary.

GWASANAETH CLYWELED / AUDIO-VISUAL AIDS SERVICE

This service is available to all affiliated associations and individual members. Further information is available from the Audio-Visual Aids Secretary.

DARLITHWYR / DANGOSWYR, LECTURERS / DEMONSTRATORS

The names and addresses of lecturers and demonstrators, recommended by associations affiliated to the WBKA, are available from the General Secretary.

CYNLLUN CYSWLLT CHWYSTRELLU / SPRAY LIAISON SCHEME

Information is available from the General secretary

✉ ☎

SIOEAU / SHOWS

Honey/beekeeping sections are included at the Royal Welsh Agricultural Show, Llanelwedd, (OS ref: SO040520) during July, and at county, town and village shows throughout Wales. Information relating to these events may be obtained from secretaries of associations in the locality of the shows.

The historic FFAIR FEL ABERCONWY is held annually in the main street of the town, (OS ref: SH278378), on 13th September. Further information is available from the secretary of Conwy Association.

RHEOLAU CYFREITHIOL / STATUTORY REGULATIONS

The administration of the statutory regulations governing all aspects of beekeeping in Wales, is the responsibility of the Wales National Assembly, Caerdydd, CF99 1NA Phone (02920) 825111 Fax: (02920) 823352 Matters concerning statutory regulations, their implications and execution, should be addressed to the Minister of Agriculture and Rural Affairs, Wales National Assembly, at the above address.

AELODAETH UNIGOL- TANYSGRIFAU/ INDIVIDUAL MEMBERSHIP SUBSCRIPTIONS

Ian Hubbuck
White Cottage
Berriew
SY21 8BB
01686 640205
ianhubbuck@hotmail.com

INSURANCE:

John Rees
insurance@wbka.com

AUDIO VISUAL AIDS:

F. G. Eckton
Cartref
Llanafan Fawr,
Llanfair ym Muallt LD2 3LT
01591 620456

CONVENTION SECRETARY:

Graham Wheeler
mertyndowning@
btinternet.com

CONVENTION TRADE STANDS SECRETARY:

Wally Shaw
Llwyn Ysgaw, Dwyran,
Llanfairpwll, Anglesey
LL61 6RH 01248 430811
waltershaw301@
btinternet.com

WBKA/CGC

CYMDEITHASAU TADOGOL A'U YSGRIFENYDDION / AFFILIATED ASSOCIATIONS AND SECRETARIES

ABERYSTWYTH, Ann Ovens,
Tan-y-Cae, Nr Talybont,
Ceredigion,
SY24 5DP
01970 832359
ann.ovens@btinternet.com
ANGLESEY, Ian Gibbs
Dryll, Bodorgan
Ynys Mon
LL62 5AD
01407 840314
secretaryabka@gmail.com
BRECKNOCK AND RADNOR,
Dr Gillian Todd, Meadow
Breeze, Llanddew, Brecon
LD3 9ST
01874610902 07971314798
gbtodd@btinternet.com
BRIDGEND, Sue Verran Ty Mel,
Maesteg Rd. Bridgend
CF32 0EE
01656 729699
verran@btinternet.com
CARDIFF AND VALE, Annie
Newsam
Stonecroft, Mountain Road,
Bedwas, Caerphilly, CF83 8ER
annienewsam@hotmail.co.uk
CARMARTHEN, Brian Jones
Cwmburry Honey Farm,
Ferryside, Carmarthenshire,
SA17 5TW
01267 267318
beegeejay2003@yahoo.co.uk

CONWY, Mr Peter McFadden,
Ynys Goch
Ty'n y Groes,
Conwy LL32 8UH
01492 650851
peter@honeyfair.freeserve.
co.uk
EAST CARMARTHEN
Linda Wallis,
Maestroyddyn Fach
Harford
Llanwrda
Carmarthenshire
SA19 8DU
01558 650774
linandbaz@aol.com
FLINT AND DISTRICT,
Jill and Graham Wheeler,
Mertyn Downing, Whitford
Holywell, Flintshire,
CH8 9EP.
01745 560557
mertyndowning@btinternet.
com
GWENYNWYR CYMRAEG
CEREDIGION   W.I.Griffiths,
Llain Deg, Comins Coch,
Aberystwyth, SY23 3BG
01970 623334
 wilmair@btinternet.com
LAMPETER AND DISTRICT
Mr Gordon Lumby,
Gwynfryn, Brynteg,
Llanybydder,
SA40 9UX
01570 480571
g.lumby@btopenworld.com

LLEYN AC EIFIONYDD
Amanda Bristow,
Bryngwydion, Pontllyfni,
Gwynedd
LL54 5EY 01286 831328
amanda.bristow@egnitec.com
MEIRIONNYDD,
Sue Townsend,
01341 430262
suetownsend@tesco.net
bazurka@aol.com
MONTGOMERYSHIRE,
Maggie Armstrong,
20 Dol-y-Felin
Abermule
Powys
SY15 6BB
01686 630447
secretary@montybees.org.uk
PEMBROKESHIRE,
Paul Eades
01437 899928
secretarypbka@hotmail.com
SOUTH CLWYD,
Mrs Carol Keys-Shaw, Y Beudy,
Maesmor Hall, Maerdy
Corwen LL21 0NS
01490 460592
c.keysshaw@btinternet.com
SWANSEA, Paul Lyons,
2 West Cliff, Southgate,
Swansea, SA3 2AN.
paul.lyons@bt.com

TEIFISIDE, Donald Adams,
Bryngwrog
Beulah
Newcastle Emlyn
Ceredigion
SA38 9QR
07932 336076
dee@orangehome.co.uk

WEST GLAMORGAN,
Mr John Beynon,
48, Whitestone Avenue,
Bishopston,
Swansea. SA3 3DA
01792 232810,
jakbeynon@btinternet.com

HEB DADOGU/NON
AFFILIATED:
Mrs J Bromley
Ty Hir, Monmouth Road
Raglan, Usk. NP15 2ET
01291 690331
bromleyjan@hotmail.com

BEIRNIAID SIOE FÊL TRWYDDEDIG / WBKA QUALIFIED HONEY SHOW

TERRY E. ASHLEY
Meadow Cottage,
11 Elton Lane, Winterley
Sandbach CW11 4TN
M. J. BADGER MBE
14 Thorn Lane, Leeds
LS8 1NN
M BESSANT
Gwili Lodge, Heol
Lotwen, Rhydaman
SA18 3RP
ROBERT BREWER
PO Box 369, Hiawassee,
Georgia, USA
TOM CANNING
151 Portadown Road,
Armagh, Co Armagh
BT61 9HL
LES CHIRNSIDE
Bryn-y-Pant Cottage,
Upper Llanover,
Abergavenny NP7 9ES

CARYS EDWARDS
Ty Cerrig, Ganllwyd,
Dolgellau LL40 2TN
IFOR C. EDWARDS
Lleifior, Pontrhydygroes,
Ystrad Meurig SY25 6DN
STEVEN GUEST
Bridge House, Hind
Heath Road, Sandbach,
CW11 3LY
HUGH MCBRIDE
11 Ballyloughan Park
Antrim BT43 5HW
LORRAINE MCBRIDE
11 Ballyloughan Park
Antrim BT43 5HW

CECIL MCMULLAN
33 Glebe Road,
Hillsborough, County
Down
LEO MCGUINESS
89 Dunlade Road, Grey
Steel BT47 4QL
GAIL ORR
64 Ballycrone Road,
Hillsborough BT26 6NH
DINAH SWEET
Graig Fawr Lodge,
Caerphilly, CF83 1NF
REDMOND WILLIAMS
Tincurry, Cahir, Co
Tipperary Eire
MICHAEL YOUNG MBE
Mileaway, Carnreagh,
Hillsborough BT26 6LJ

NBU

✉ ☎

NATIONAL BEE UNIT,
THE FOOD AND ENVIRONMENT
RESEARCH AGENCY

www.nationalbeeunit.com

National Bee Unit
The Food and Environment
Research Agency
Sand Hutton, York, YO41
1LZ, UK

Tel.No: 01904 462510
Fax.No: 01904 462240
E-Mail: nbu@fera.gsi.gov.uk
Website:
www.nationalbeeunit.com
www.fera.defra.gov.uk
Policy:
www.defra.gov.uk

NATIONAL BEE UNIT

The National Bee Unit (NBU) is part of the executive agency of the Department for Environment, Food and Rural Affairs (Defra), and is based just outside York. The Unit is an element of Fera's Inspectorate Programme and its work covers all aspects of bee health and husbandry in England and Wales, on behalf of Defra in England and for the Welsh Government in Wales. The work of the unit includes disease and pest diagnosis, research into bee health matters, development of contingency plans for emerging threats, import risk analysis, related extension work and consultancy services to both government and industry.

BEE HEALTH INSPECTION SERVICE

NATIONAL BEE UNIT TECHNICAL
STAFF, HEAD OF UNIT
Mike Brown

NATIONAL BEE INSPECTOR
Andy Wattam
01522 789726
07775 027524

The Integrated Bee Health Programme is run by the NBU on behalf of core policy customers. The NBU has a long track record in bee husbandry and bee disease control (since 1946) and has been directly responsible for the bee inspection services in England and Wales since 1994. The NBU consists of a home-based inspectorate team, and the laboratory diagnostic and research team based at Fera, York. In addition colleagues across Fera contribute to the programme and research projects. The Bee Health Inspectorate The inspectorate team consists of approximately 50 home-based members of staff. It is headed by the National Bee Inspector (NBI), whose role it is to manage the statutory disease control and training programmes. The NBI has management responsibility for eight home-based Regional Bee Inspectors (RBIs), one heading each of the seven regions in England

and one covering Wales. The RBI in turn manages a number of Seasonal Bee Inspectors (SBIs). The RBIs and SBIs organise inspections under EU and UK legislation, submit suspect samples for diagnosis, treat colonies for foul brood and train beekeepers in bee husbandry for better disease control and greater self-sufficiency. In addition the bee inspectors also collect honey samples for residue analysis under the Statutory Honey collection agreement with Defra Veterinary Medicines Directorate (VMD). With *Aethina tumida* (Small hive beetle (SHB)) and *Tropilaelaps spp.* both notifiable under UK and EU law inspectors also undertake surveillance for these exotics in "at risk apiaries" close to identified high risk areas.

BEE DISEASE DIAGNOSTIC TEAM

The NBU's diagnostic team provides a rapid, modern service for both the inspection team and beekeepers.

The NBU laboratory is Good Laboratory Practice (GLP) compliant, a quality accreditation scheme administered by the Department of Health. All diagnostic tests are conducted according to the OIE (Office International des Epizooties) Manual of Standard Diagnostic Tests and Vaccines. The OIE is the world organisation for animal health and produce internationally recognised disease diagnosis guidelines (http//www.oie.int.) Across Fera diagnostic support is provided from teams of microbiologists acarologists, insect virologists and molecular specialists in the Fera Molecular Technology Unit (MTU).

BEES AND THE LAW

The Bees Act 1980 UK empowers Agriculture Ministers to make Orders to control pests and diseases affecting bees, and provides powers of entry for authorised persons. Under the Bees Act, The Bee Diseases and Pests Control Order 2006 for England and Wales, (there is similar legislation for Scotland and Northern Ireland) designates American foulbrood (AFB), European foulbrood (EFB), *A. tumida* (SHB) and *Tropilaelaps* mites (all species) as notifiable pests and

REGIONAL BEE INSPECTORS

Ian Molyneux
Northern Region
01204 381186
07775 119442

Charles Millar
Western Region
01694 722419
07775 119476

Nigel Semmence
Southern Region
01264 338694
07776 493649

Alan Byham
South East Region
01306 611016
07775 119447

Simon Jones
South West Region
01823 442228
07775 119459

Keith Morgan
Eastern Region
01485 520838
07919 004215

Ivor Flatman
North East Region
01924 252795
07775 119436

Frank Gellatly
Wales
01558 650663
07775 119480

FOR DETAILS OF SEASONAL BEE INSPECTORS DETAILS CONTACT THE RELEVANT RBI OR CHECK BEEBASE

NBU

✉ ☎

LABORATORY BASED STAFF
Research Co-ordinator
Giles Budge

Bee Research
Gay Marris

Laboratory Manager
Ruth Grant
Ilex Whiting
Victoria Tomkies

Apiary Manager
Damian Cierniak
Jack Wilford

Technical Advisor
Jason Learner

Administrative
Programme Support
Kate Parker,
Lesley Debenham &
Jenna Cook

defines the action which may be taken in the event of outbreaks. At the European level, the Directive on animal health requirements for trade in bees is called the Balai Directive (92/65/EEC) implemented in the UK under the Animal and Animal Products (Import and Export) Regulations. It lists American foul brood (AFB), the small hive beetle (*A. tumida*) and *Tropilaelaps* mites as notifiable pests and diseases throughout the EU (at the time of writing time neither the small hive beetle nor *Tropilaelaps* have been confirmed in Europe).

THE IMPORTATION OF BEES
It is legal to import Queen bees from third countries, the rules governing this are set out in Commission Decision 2003/881/EC, as amended by Commission Decision 2005/60/EC. The list of countries is currently restricted to, Argentina, Australia and New Zealand. It is legal to import bees freely from the EU (including queens, packages and colonies). Under the Balai directive consignments of bees moved between Member States must be accompanied by an original health certificate confirming freedom from notifiable pests and diseases. For full details on the importation of bees from within the EU or from Third countries please either consult the Defra website, BeeBase or contact the NBU.

AMERICAN AND EUROPEAN FOUL BROOD
Foul brood-infected apiaries are placed under standstill notice, supervised by the bee inspector, until the disease is cleared from the apiary and the honey from antibiotic-treated colonies is safe to harvest. We always aim to minimise the impact of this as far as possible, in co-operation with the beekeeper.

VARROA
As part of the NBU's routine field screening programme the first known case of pyrethroid resistant *Varroa* mites in the UK was discovered in apiaries in Devon in August 2001. The NBU

undertook a resistance-monitoring programme throughout England and Wales. Pyrethroid resistant *Varroa* mites are now widespread in England and Wales. To access current advice on *Varroa* and it's Management please visit BeeBase.

ADULT BEE DISEASES

The NBU also look for adult bee diseases and parasites such as Nosema species (*Nosema apis* and *Nosema ceranae*, amoeba (*Malpighamoeba mellificae*) and tracheal mites (*Acarine* or *Acarapis woodi*) from samples submitted by beekeepers. As these diseases are non-statutory this service is chargeable. For the current cost please contact the NBU or visit the website.. Bees that have been imported from designated Third countries are also checked for disease and are also screened for exotic pests potentially harmful to UK beekeeping.

EXOTICS

Beekeepers must make themselves aware of the potential threats to beekeeping in the UK. The field inspection team monitors for potential exotics, the SHB and *Tropilaelaps spp*. The laboratory team also routinely screen import samples and suspect samples submitted for identification by both beekeepers and the field team.

PESTICIDE MONITORING

The Wildlife Incident Investigation Scheme (WIIS) is a unique scheme for monitoring the effects of pesticides on wildlife, including beneficial invertebrates such as honey bees. It is led by the Chemicals Regulation Directorate (CRD) with Natural England managing and undertaking site enquiries on their behalf; The Food and Environment Research Agency (Fera) carry out disease and pesticide analysis and, if appropriate, the Veterinary Laboratories Agency (VLA) carry out post mortems on wildlife. Information gathered is fed into the approval process for pesticides and helps in the verification and improvement of pesticide risk assessments.

It can also result in changes to label recommendations on pesticide products. It is not provided as a personal service to beekeepers wishing to seek evidence for the

purpose of civil litigation but can lead to enforcement action being taken by the enforcer if the misuse or abuse of a product is identified as part of this process. For more information please see the website.

RESEARCH & DEVELOPMENT
A programme of research and development within the group underpins the Unit's work. They also have long-established links with many European and world wide research centres, universities and the beekeeping industry. The primary aim of our R&D is to improve our understanding of the issues which impact bee health. The NBU also actively supports PhD students, some of which are funded using donations from the beekeeping industry.
For an update on the current R&D work of the unit please see BeeBase.

RISK ASSESSMENT
The National Bee Unit manages 150 honey bee colonies and has much experience in assessing the effects and efficacy of veterinary bee medicines (e.g., varroacides) and pesticides in both field and laboratory tests. Our Good Laboratory Practice (GLP) accreditation allows us to undertake a wide range of routine and specially designed laboratory, semi-field and field studies on honeybees and bumblebees for regulatory authorities and industry worldwide.

EXTENSION
The NBU trains beekeepers in several ways: local courses and advisory visits run by the inspectors, and national courses held at the York laboratory. The NBU annually hosts the National Diploma in Beekeeping residential courses and has also been host to visiting overseas workers and researchers. NBU York based staff also provide training to beekeepers at local and regional beekeeper meetings.

HEALTHY BEES PLAN

The Healthy Bees Plan was published by Defra and the Welsh Assembly Government in March 2009 following consultation with beekeepers and the main Beekeeping Associations. It sets out a plan for Government, beekeepers and other stakeholders to work together to respond effectively to pest and disease threats and to put in place programmes to ensure a sustainable and productive future for beekeeping In England and Wales. The Healthy Bees Plan consists of three working groups that report to the project management board to help deliver the five major objectives of the plan. To view the Healthy Bees Plan, please see the website.

(This is the most recent information received from the National Bee Unit). BeeBase is the National Bee Unit website. It is designed for beekeepers and supports Defra, WAG and Scotland's Bee Health Programmes and the Healthy Bees Plan, which set out to protect and sustain our valuable national bee stocks. Our website provides a wide range of free information for beekeepers, to help keep their honey bees healthy. We hope both new and experienced beekeepers will find this an extremely useful resource and sign up to BeeBase. Knowing the distribution of beekeepers and their apiaries across the country helps us to effectively monitor and control the spread of serious honey bee pests and diseases, as well as provide up-to-date information in keeping bees healthy and productive. By telling us who you are you'll be playing a very important part in helping to maintain and sustain honey bees for the future. To register as a beekeeper please visit BeeBase.

DARD

✉ ☎

DEPARTMENT OF AGRICULTURE AND RURAL DEVELOPMENT

WWW.dardni.gov.uk

BEE DISEASE DIAGNOSTICS:
Sam Clawson
Agri-Food and Biosciences Institute (AFBI)
Newforge Lane
BELFAST BT9 5PX
Tel: 028 9025 5289
Email: Sam.Clawson@ afbini.gov.uk

TRAINING COURSES:
Jennifer Ball
Greenmount Campus
College of Agriculture Food and Rural Enterprise:
Information is available from the College at
Tel: 028 9442 6879
Text phone: 028 9052 4420
Email: Jennifer.Ball@ dardni.gov.uk

BEE INSPECTIONS:
Thomas Williamson
Agri-food Inspection Branch,
DARD, Glenree House,
Carnbane Industrial Estate,
Newry, Co Down, BT35 6EF
Tel: 028 3889 2374
Fax: 028 3025 3255
Email: Thomas.Williamson@ dardni.gov.uk

Honeybee Regional Report for Northern Ireland 2012

Bee Health Surveys

A questionnaire survey for Bee Husbandry issues has been circulated annually to beekeepers via beekeeping associations since 2009. The results of the 2011 survey are available as a pdf on the AFBI website (www.afbini.gov.uk). This showed colony losses for 2011 were 16% compared to 13% in 2010. Seventy percent of beekeepers reported no losses. The 2012 survey results are currently being processed but will be available on the AFBI website in November.

Bee Health Inspections

The Bee Inspectorate carried out surveys for American foul brood, European foul brood, Small Hive beetle and Tropilaelaps mite along with resistance testing of varroa mites to pyrethroids. American foul brood remains a problem for beekeepers in Northern Ireland with 10 apiaries found to have the disease by early September compared to twelve apiaries in total for 2011. Inspections were also carried out for European foul brood without any incidents recorded. Surveys continued for Small Hive Beetle and Tropilaelaps mite. Apiaries in the vicinity of ports or fruit importers were targeted for Small Hive Beetle inspections using corriboard shelter traps, while apiaries that had imported in the past were selected and hive scrapings examined for Tropilaelaps mite.

Varroa
Varroa is ubiquitous in Northern Ireland, consequently no systematic studies of prevalence are conducted. Samples continue to be submitted to monitor varroacide resistance. Six of eleven samples suitable for testing in 2012 returned a positive result for varroacide resistance.

Adult Bee Disease Diagnostics
Nosema ceranae was first recorded in Northern Ireland in 2010. *N. ceranae* is an emergent pathogen of western honeybees. It is similar to the endemic species, *Nosema apis* but is considered to produce a more virulent disease than *N. apis,* probably reflecting its more recent association with the western honeybee. Samples submitted and positive for *Nosema* are screened for *N. ceranae* on an ad-hoc basis. Since April 2011, 142 samples were tested for *N. apis* and *N. ceranae* with 78 (55%) positive for a *Nosema* infection. Of these, 41 (53%) contained *N. ceranae*. Note, however, this should not be used as an indicator of prevalence, as samples were not spatially representative of colony distribution in Northern Ireland.

Up to September 2012, 109 samples of bees have been submitted to the laboratory for disease diagnosis, 33 were positive microscopically for *Nosema* and 15 positive for acarine. In 2011, we had a total of 93 similar samples of which 38 proved positive microscopically for *Nosema* and 13 positive for acarine disease.

Residue Sampling
Honey samples were again lifted this year for testing for residues of veterinary medicines and environmental contaminants. Samples lifted last year were found to be satisfactory.

Imports
Twelve direct Queen imports were notified to DARD in 2012 from Greece and Cyprus. Notification levels have been much less than previous

years. Follow-up inspections were carried out to check records and health certificates for notified imports.

The Bee Diseases and Pests Control Order (Northern Ireland) 2007
The above Order came into operation on the 21 May 2007, which brought our list of notifiable pests and diseases into line with England and Wales.

Bee Health Contingency Plan
The Bee Health Contingency Plan is reviewed annually and an updated version was published on the DARD Internet in September 2012.

Strategy for the Sustainability of the Honey Bee
The Strategy for the Sustainability of the Honey Bee was published in February 2011 and aims to achieve a sustainable and healthy population of honey bees for both pollination and honey production in the north of Ireland through strengthened partnership working between Government and Stakeholders. The Strategy confirms DARD's ongoing commitment to help protect and improve the health of honey bees and support the sector in its efforts to sustain and support beekeeping. The Ulster Beekeepers Association (UBKA) and the Institute of NI Beekeepers (INIB) have made a commitment to support the Strategy intentions. The Strategy is aimed at both policy makers and beekeepers, and importantly, identifies the roles and responsibilities of the different stakeholders in delivering its aims and outcomes. It seeks to address the current challenges facing beekeepers and provides a plan of action aimed at sustaining the health of honey bees and beekeeping in the north of Ireland for the next decade.

The Strategy for the Sustainability of the Honey Bee can be viewed at:
http://www.dardni.gov.uk/index/publications/pubs-dard-fisheries-farming-and-food/publications-dard-strategy-for-the-sustainability-of-the-honey-bee.htm

DARD

Jim Crummie
Head of Crop Certification Plant & Bee Health Inspectorate, DARD
Thomas Williamson
Senior Crop Certification Plant & Bee Health Inspector, DARD
Sam Clawson
Bee Disease Diagnostics, AFBI
Seamus Hughes
Farm Policy Branch, DARD

SG-AFRC

The Scottish
Government

THE SCOTTISH GOVERNMENT AGRICULTURE, FOOD AND RURAL COMMUNITIES DIRECTORATE (AFRC) - RURAL PAYMENTS AND INSPECTIONS DIRECTORATE (RPID)

HEADQUARTERS
Lead Bee Inspector
Stephen Sunderland,
P Spur, Saughton House,
Broomhouse Drive,
Edinburgh, EH11 3XD
Tel: 0300 244 6672
e-mail: beesmailbox@
scotland.gsi.gov.uk

The Scottish Government (SG) is responsible for bee health Policy in Scotland. SG recognises the importance of a strong Bee health programme, not only for the production of honey, but also for the contribution that bees make to the pollination of many crop species and to the wider environment.

Honey bees are susceptible to a variety of threats, including pests and diseases, the likelihood and consequences of which have increased significantly over the last few years.

The Scottish Government takes very seriously any biosecurity threat to the sustainability of the apiculture sector and is working closely with colleagues in Food and Environment Research Agency's (Fera) National Bee Unit (NBU) to enable a more joined up approach to be taken throughout Great Britain on the issues surrounding bee health.

The Scottish Government has invested in the NBU's National web-based database for beekeepers "BeeBase" and actively encourages beekeepers to register onto the system. This service will provide bee health and disease outbreak information and will also assist Bee Inspectors in disease control. BeeBase also provides information on legislation, pests and disease recognition and control, interactive maps, current research areas and key contacts.

Beekeepers have a significant role to play in ensuring disease management and control within their own apiaries are in order as they have a legal obligation to report any suspicion of a notifiable disease or pest to the Bee Inspector at their local SGRPID Area Office. Bee Inspectors are responsible for the operation of The Bee Diseases and Pests Control (Scotland) Order 2007 in their area with duties including:-

- Inspection of apiaries for presence of statutory bee diseases
- Taking and delivering samples to SASA
- Issuing and removal of 'Standstill Notices'
- Issuing of 'Destruction Notices' and supervising destruction
- Informing beekeepers of treatment options for European Foul Brood (EFB), where appropriate
- Granting the option, after taking account of the recommendations of SASA, and carrying out treatment
- Carrying out follow-up inspections after destruction or treatment

SASA

- **Science and Advice for Scottish Agriculture (SASA)** is responsible for providing specialist technical support where duties include:
- Examination of submitted samples suspected of being infected with American Foul Brood, European Foul Brood, Small Hive Beetle (SHB) or *Tropilaelaps*.
- Reporting results on which pathogen or pest is present
- Recommending, in consultation with the Bee Inspector, the most suitable option, destruction or treatment, for each individual case of EFB.
- Where treatment is agreed, ordering supplies of the approved antibiotic
- Provision of a free diagnostic service to beekeepers to identify and confirm the presence of varroa.
- Maintaining technical liaison with NBU and providing technical documentation as required
- Providing training courses and demonstration material as required

SASA (SCIENCE AND ADVICE FOR SCOTTISH AGRICULTURE)
1 Roddinglaw Rd,
Edinburgh, EH12 9FJ

BEE DISEASES,
FIONA HIGHET
Plant Health Section
(0131) 244 8817

PESTICIDE INCIDENTS,
ELIZABETH SHARP
Chemistry Section
(0131) 244 8874

SG-AFRC
✉ ☎

PESTICIDE INCIDENTS

As part of the Wildlife Incident Investigation Scheme (WIIS), SASA undertakes analytical investigations into bee mortalities where pesticide poisoning may have been involved. Beekeepers should send samples of dead bees (200) direct to SASA, Chemistry Section, for analysis. In the case of major incidents, beekeepers are advised to contact their nearest SGRPID Area Office so that an early field investigation can be instigated.

THE FOLLOWING SCOTTISH GOVERNMENT RURAL PAYMENTS AND INSPECTIONS DIRECTORATE (SGRPID) STAFF ARE AUTHORISED BEE INSPECTORS. ALL BEE INSPECTORS HAVE EMAIL ADDRESSES AS "FIRSTNAME.SURNAME@SCOTLAND.GSI.GOV.UK"

EDINBURGH (HQ)
Steve Sunderland
(Lead Bee Inspector)
P Spur, Saughton House,
Broomhouse Drive,
Edinburgh, EH11 3XD
Tel: 0300 244 6672
Fax: 0300 244 9797

GRAMPIAN (INVERURIE AREA OFFICE)
Kirsteen Sutherland
Thainstone Court,
Inverurie, Grampian,
Aberdeenshire, AB51 5YA
Tel: (01467) 626247
Fax: (01467) 626217

SOUTHERN (DUMFRIES AREA OFFICE)
Angus Cameron
161 Brooms Road
Dumfries, DG1 3ES
Tel: (01387) 274400
Fax: (01387) 274440

CENTRAL (PERTH AREA OFFICE)
Kelly Callwood
Strathearn House
Broxden Business Park
Lamberkine Drive
Perth, PH1 1RZ
Tel: 01738 602043

HIGHLAND (INVERNESS AREA OFFICE)
Clem Cuthbert
Longman House
28 Longman Road
Inverness, IV1 1SF
Tel: 01463 253 053

SOUTH EASTERN (GALASHIELS AREA OFFICE)
Angus MacAskill
Cotgreen Road
Tweedbank, Galashiels
Scottish Borders, TD1 3SG
Tel: (01896) 892400
Fax: (01896) 892424

SOUTH WESTERN (AYR AREA OFFICE)
John Smith
Russell House
King Street
Ayr
South Ayrshire
KA8 0BG
Tel: (01292) 291300
Fax: (01292) 291301

SCOTLAND'S RURAL COLLEGE (SRUC)

The Scottish Government supports a full-time apiculture specialist (Graeme Sharpe) who provides comprehensive advisory, training and education programmes for beekeepers throughout Scotland on all aspects of Integrated Pest Management, good husbandry (including the control of Varroa) and management practices. SAC also promotes the awareness of notifiable bee diseases and pests.

GRAEME SHARPE, APICULTURE SPECIALIST,
SAC Consulting, Scotland's Rural College (SRUC), Veterinary Services John Niven Building Auchincruive Estate Ayr, Ayrshire KA6 5HW
Tel:01292 525375

WWW.SCOTLAND.GOV.UK/TOPICS/APICULTURE/GRANTS/INSPECTIONS/BEEINSPECTIONS

USEFUL TABLES

BEEKEEPING METRIC CONVERTION TABLES

°CENT	FAHR	INCH	MM	INCH	MM	INCH	MM
0	32	$^1/_{25}$	1	$1^5/_8$	42	10	254
5	40	$^1/_{12}$	2	$1^{11}/_{16}$	43	$10^1/_4$	260
7	44	$^1/_8$	3	$1^9/_{20}$	48	$11^1/_4$	286
30	86	$^1/_{16}$	5	2	51	$11^1/_2$	292
34	92	$^1/_4$	6	3	76	$11^3/_4$	298
38	100	$^5/_{16}$	8	$4^1/_4$	108	12	305
43	110	$^3/_8$	9	$4^1/_2$	114	14	356
49	120	$^1/_2$	12.5	$4^3/_4$	121	$16^1/_4$	413
54	130	$^5/_8$	16	$5^1/_2$	140	$16^1/_2$	49
60	140	$^3/_4$	18	$5^3/_4$	146	17	431
62	144	$^7/_8$	22	6	152	$17^5/_8$	448
82	180	1	25	$6^1/_4$	159	$18^1/_8$	460
90	194	$1^1/_{16}$	27	$8^1/_4$	216	$18^1/_4$	483
100	212	$1^3/_8$	35	$8^3/_4$	223	20	508
		$1^9/_{20}$	37	$9^1/_8$	232	$21^1/_2$	546
		$1^1/_2$	38	$9^3/_8$	239	$21^3/_4$	552
				$9^9/_{16}$	246	22	559

INTERNATIONAL QUEEN MARKING COLOURS

YEAR ENDING	COLOUR	REMEMBER
1 & 6	WHITE	Will
2 & 7	YELLOW	You
3 & 8	RED	Raise
4 & 9	GREEN	Good
5 & 0	BLUE	Bees?

USEFUL TABLES

BOTTOM BEE-SPACE HIVES

No, of cells in brood box
Lug length (MM)
Frame spacing (mm)
Frame size (mm)
No. frames
Hive tvpe

Hive type		No. frames	Frame size (mm)	Frame spacing (mm)	Lug length (MM)	No. of cells in brood box
National	BROOD	11	356 x 216	37	38	58000
	SUPER	10	356 x 140	42	38	36000
Modified Commercial	BROOD	11	406 x 254	37	16	75000
	SUPER	10	406 x 152	42	16	

TOP BEE-SPACE HIVES

No, of cells in brood box
Lug length (MM)
Frame spacing (mm)
Frame size (mm)
No. frames
Hive tvpe

Hive type		No. frames	Frame size (mm)	Frame spacing (mm)	Lug length (MM)	No. of cells in brood box
Smith	BROOD	11	356 x 216	37	18	58000
	SUPER	10	356 x 140	42	18	36000
Langstroth	BROOD	10	448 x 232	35	16	68000
	SUPER	10	448 x 140	35	16	
Jumbo	BROOD	10	448 x 286	35	16	85000
	SUPER	10	448 x 140	35	16	
Modified Dadant	BROOD	11	448 x 286	37	16	93000
	SUPER	10	448 x 159	42	16	

USEFUL TABLES

CONVERSION FACTORS

TEMPERATURE

Fahrenheit > Celcius (Centigrade)	- 32, x 0.5555 ($^5/_9$)
Celcius > Fahrenheit	x 1.8 ($^9/_5$), + 32

WEIGHT

Ounces > Pounds	x 28.3495
Pounds > Grams	x 453.59237
Hundredweights > Kilograms	x 50.8
Grams > Ounces	'/. 28.3495
Kilograms > Pounds	x 2.2142

LENGTH

Inches > Centimetres	x 2.54
Yards > Metres	x 0.9144
Miles > Kilometres	x 1.609
Centimetres > Inches	x 0.3937
Metres > Yards	x 1.0936
Kilometres > Miles	'/. 1.609

AREA

Acres > Hectares	x 0.404686
Hectares > Acres	'/. 2.47105

VOLUME

Pints > Litres	x 0.5683
Gallons > Litres	x 4.546
Litres > Pints	x 1.7598
Litres > Gallons	x 0.21997

WORD SEARCH ANSWERS

THOMAS JEFFERSON
SYLVIA PLATH
VIKTOR YUSCHENKO
MARIA VON TRAPP
LEO TOLSTOY
HENRY FONDA
BENJAMIN FRANKLIN
LORD BADEN POWELL
FRANCOIS MITTERRAND
PYTHAGORAS
SIR EDMUND HILLARY
ELWYN BROOKS WHITE
GEORGE WASHINGTON
THOMAS EDISON
GREGOR MENDEL
JOHN WHITTIER
PAUL THEROUX
DEMOCRITUS
LE QUY QUYNH
PETER FONDA
VIRGIL

www.ingramcontent.com/pod-product-compliance
Lightning Source LLC
Chambersburg PA
CBHW072136270326
41931CB00010B/1777